CARRIAGES & COACHES

CARRIAGES & COACHES

RALPH STRAUS

ISBN: 978-93-5420-515-6

Published: 1912

LECTOR HOUSE LLP
E-MAIL: lectorpublishing@gmail.com

The State Coach of Great Britain

CARRIAGES & COACHES
THEIR HISTORY & THEIR EVOLUTION

BY

RALPH STRAUS

**FULLY ILLUSTRATED WITH REPRODUCTIONS
FROM OLD PRINTS, CONTEMPORARY
DRAWINGS & PHOTOGRAPHS**

MCMXII

TO
B. S. S.

PREFACE

I AM not a coachbuilder. Though such a pronouncement will seem entirely superfluous to any coachbuilder who reads the following pages, it is not perhaps a wholly unnecessary remark. For, with one or two exceptions, such books upon the evolution or structure of vehicles as have been written have been the work of industrious coachbuilders. And I have not the least doubt that they are eminently the fit and proper folk to carry out any such task. It is a melancholy fact, however, that useful though these books may be to coachbuilders, they lack, again with one or two exceptions, any general interest to the layman. The language in which they are written is, to say the least, peculiar, and the authors have obviously had small training in the art of book-making. On the other hand, there is a whole library of books dealing with the old stage and mail coaches, with all the romance and adventure of the roads, packed with delightful anecdotes and personal reminiscences. But such books hardly touch upon the structure of the coaches themselves, and, so far as I know, there is no book entirely devoted to a non-technical description of carriages in general, based upon a chronological arrangement.

The nearest approach to such a book is Mr. G. A. Thrupp's *The History of Coaches*, published in 1877, a meritorious undertaking from which I have freely quoted. Here, however, there are numerous gaps which I have endeavoured to fill, and the various lectures from which it was composed do not fit together so aptly as might be. As a whole, it is diffuse. Sir Walter Gilbey's two books, *Early Carriages and Roads* and *Modern Carriages*, have also been of great assistance, but here, too, the ground covered is not so large as in the following pages. Other pamphlets and small books have appeared in this country, but seemingly owe a great deal of their information to Mr. Thrupp's work. Indeed, I notice that some of the authors have been almost criminally forgetful of their inverted commas. For purely technical details there are, of course, many books and trade papers to consult; but with these I have not been concerned.

In the present book there are, indeed, large gaps, and it is not to be taken either as a manual of the art of coach-building or as a history of locomotion. It is merely a book about carriages, in which particular regard has been paid to chronological sequence, and particular attention to such individual carriages as have at all withstood the test of social history. And it is written by a layman who, until he enquired into the subject, had never looked at a carriage with any particular emotion. The result of his labours, therefore, is not meant for the expert, but for the general reader, who may have pondered over the various vehicles he has seen, and idly wondered how they may have been evolved.

Where possible, I have endeavoured to quote from contemporary authors and documents. Most of such quotations are now included in a carriage book for the first time.

I wish to thank the various publishers and authors who have given me permission to reprint illustrations of carriages in books published or written by them. Also I am obliged to Messrs. Maggs Bros., the well-known booksellers, for permission to photograph a rare print entitled *The Carriage Match*, in their possession.

RALPH STRAUS.

BADMINTON CLUB, *August, 1912.*

CONTENTS

LIST OF ILLUSTRATIONS

CHAPTER THE FIRST
THE PRIMITIVE VEHICLE

"This is a traveller, sir, knows men and
Manners, and has plough'd up sea so far,
Till both the poles have knock'd; has seen the sun
Take coach, and can distinguish the colour
Of his horses, and their kinds."

Beaumont and Fletcher.

IT has been suggested that although in a generality of cases nature has fore-stalled the ingenious mechanician, man for his wheel has had to evolve an apparatus which has no counterpart in his primitive environment—in other words, that there is nothing in nature which corresponds to the *wheel*. Yet even the most superficial inquiry into the nature of the earliest vehicles must do much to refute such a suggestion. Primitive wheels were simply thick logs cut from a tree-trunk, probably for firewood. At some time or another these logs must have rolled of their own accord from a higher to a lower piece of ground, and from man's observation of this simple phenomenon must have come the first idea of a wheel. If a round object could roll of its own accord, it could also be made to roll.

Yet it is to be noticed that the earliest methods of locomotion, other than those purely muscular, such as walking and riding, knew nothing of wheels. Such methods depended primarily upon the enormously significant discovery that a man could drag a heavier weight than he could carry, and what applied to a man also applied to a beast. Possibly such discovery followed on the mere observation of objects being carried down the stream of some river, and perhaps a rudely constructed raft should be considered to be the earliest form of vehicle. From the raft proper to a raft to be used upon land was but a step, and the first land vehicle, whenever or wherever it was made, assuredly took a form which to this day is in common use in some countries. This was the sledge. On a sledge heavy loads could be dragged over the ground, and experience sooner or later must have shown what was the best form of apparatus for such work. As so often happens, moreover, in mechanical contrivances, the earliest sledge of which there is record—a sculptured representation in an Egyptian temple—bears a remarkable resemblance to those in use at the present time.[1] Then, as now, men used two long runners with upturned

---

[1] "In Europe, sledge is the name applied to a low kind of cart, but in America the

ends in front and cross-pieces to unite them and bear the load. Such sledges were largely used to convey the huge stones with which the Egyptians raised their solemn masses of masonry and, incidentally, also as a hearse. In time, however, it was found that better results were obtained by the use of another and rather more complicated apparatus which had for its chief component—a wheel. This second discovery that to roll a burden proved an easier task than to drag it was fraught with such tremendous consequences as altered the entire history of the world.

It remained to find a better fulcrum than that afforded by the rough turf over which such logs, when burdened, were rolled. What probably followed is well described by Bridges Adams.[2] "The next process," he thinks, "would naturally be that of cutting a hole through the roller in which to insert the lever. The convenience of several holes in the circumference of the roller would then become apparent, and there would be formed an embryo wheel nave. It could not fail to be remarked also, that the larger the roller, the greater the facility for turning it, and consequently the greater the load that could be borne upon it." Owing to the difficulty of using such large logs, he goes on to suggest, a time would come when it was found that a roller need not bear upon the ground throughout its length, but only at its extremities. So from the single roller would be evolved two rough wheels joined by a beam, square at first though afterwards rounded, upon which could be fixed a frame for the load.

Such axle and wheels would revolve together and keep the required position by means of pieces of wood which may be compared with the thole-pins of a boat. And it is a remarkable fact that until last century such primitive carts were in use in Portugal and parts of South America. The chief drawback to a vehicle of this kind is its inability to turn in a small space, and the pioneers, whoever they were, finally discovered the principle of the fixed axle-tree, the wheels revolving upon their own centre. So, "instead of fixing the cross-beam or axle in a square hole," these pioneers "would contrive it to play easily in a round one of a conical form, that being the easiest form of adjustment." Such a car as this, with solid wheels and a rude frame, was used by the Romans, and is still to be seen in parts of Chili. The next process in the evolution of the wheel doubtless followed upon the necessity of economising with large sections of wood, and there was finally invented a wheel made of three portions—a central pierced part, the nave, an outside circular piece, the rim or felloe, and two or more cross-pieces, joining the two, the spokes. Of these the felloes would tend to wear soonest, and a double set would be applied to the spokes, as was the case until recently in the ox-carts of the Pampas, or *barcos de tierra*, as they were called by the natives.

And indeed, the first carriages of which we have particular information, the chariots of the Egyptians and their neighbours, differ essentially from such primitive carts only in the delicacy and ornamentation of the carriage body.

word has been abbreviated to sled or changed to sleigh, which in either case involves the idea that a *sliding* vehicle is meant. In the rural districts, the farmer employs a machine we call a stone-sledge. This is commonly made from a plank, the flat under surface of which is forced along the surface of the ground by ox-power." *The World on Wheels*. Ezra N. Stratton. New York, 1888.

[2] *English Pleasure Carriages*. By William Bridges Adams. London, 1837.

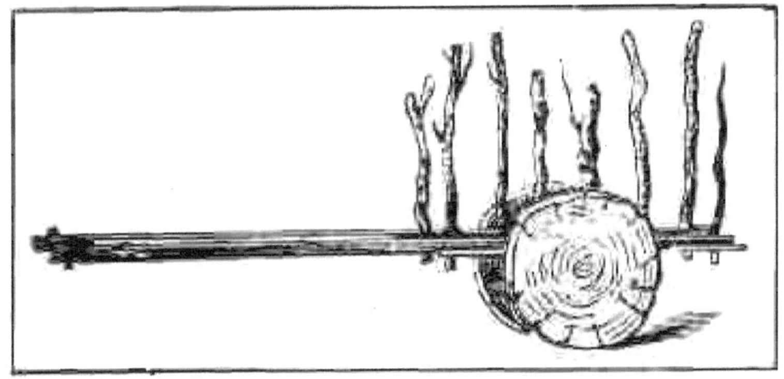

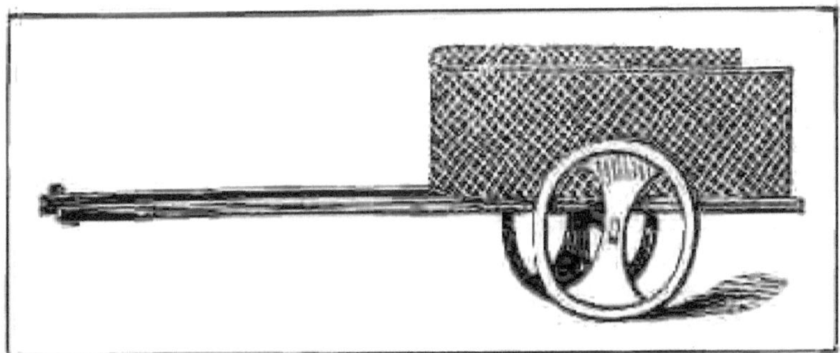

Types of Primitive Carts

Various vehicles are mentioned in the Bible, though one must be chary of differentiating between them merely because the translators have given them different names. Both waggons and chariots are mentioned in Genesis. Jacob's family were sent to him in a waggon. Joseph rode in the second chariot of Pharaoh as a particular mark of favour. At the time of the Exodus, war-chariots formed an important part of the Egyptian army, and indeed, right through the various dynasties, there is an almost continuous mention of their use.[3] "The deft craftsmen of Egypt," says Breasted,[4] "soon mastered the art of chariot-making, and the stables of the Pharaoh contained thousands of the best horses to be had in Asia." About 1500 B.C. Thutmose III went forth to battle in "a glittering chariot of electrum." He slew the enemy's leader, and took captive their princes and "their chariots, wrought with gold, bound to their horses." These barbarians also had "chariots of silver," though this probably means that they were built of wood and strengthened or decorated with silver. At the dissolution of the Empire the Hittites had increased wonderfully in power, and it is told of them that they excelled all other nations in the art of chariotry. The Hittite chariot was larger and more heavily built than that of the Egyptians, as it bore three men, driver, bowman, and shield-bearer, while the Egyptian was satisfied with two. The enormous number of chariots used in warfare is shown by the fact that in the fourteenth century before Christ, when the Egyptians defeated the Syrians at Megiddo, nearly a thousand were captured, and against Ramses II the Hittites put no less than 2500 into the field.

"The Egyptian chariots," says H. A. White,[5] "were of light and simple construction, the material employed being wood, as is proved by sculptures representing the manufacture of chariots. The axle was set far back, and the bottom of the car, which rested on this and on the pole, was sometimes formed of a frame interlaced with a network of thongs or ropes. The chariot was entirely open behind and for the greater part of the sides, which were formed by a curved rail rising from each side of the back of the base, and resting on a wooden upright above the pole in front. From this rail, which was strengthened by leather thongs, a bowcase of leather, often richly ornamented, hung on the right-hand side, slanting forwards; while the quiver and spear cases inclined in the opposite direction. The wheels, which were fastened on the axle by a linch-pin secured with a short thong, had six spokes in the case of war chariots, but in private vehicles sometimes only four.[6] The pole sloped upwards, and to the end of it a curved yoke was attached. A small saddle at each end of the yoke rest-

[3] "They also possessed baggage-carts shaped like the chariots. One of these appears to have had a very high, six-spoked wheel and a curved roof box. In front of the box is a low seat, from underneath which projects a crooked drag-pole." *Stratton.*

[4] *A History of Egypt.* J. H. Breasted. New York. 1909.

[5] *Dictionary of the Bible.* 1906. Edited by J. Hastings. Art. *Chariot.*

[6] "We account for this difference by supposing that in battle, when success depended in a great measure upon the stability of the chariot, special care was taken to provide a strong wheel, while a weaker one was considered good enough for a more peaceful employment, a four-spoked wheel in those days being much cheaper and lighter." *Stratton.*

ed on the withers of the horses, and was secured in its place by breast-band and girth. No traces are to be seen. The bridle was often ornamented; a bearing-rein was fastened to the saddle, and the other reins passed through a ring at the side of this. The number of horses to a chariot seems always to have been two; and in the car, which contained no seat, only rarely are more than two persons depicted, except in triumphal processions.

Assyrian Chariot

(From Smith's "Concise History of English Carriages")

"Assyrian chariots did not differ in any essential points from the Egyptian.[7] They were, however, completely panelled at the sides, and a shield was sometimes hung at the back. The wheels had six, or, at a later period, eight spokes; the felloes were broad, and seem to have been formed of three distinct circles of wood,

---

[7] The Assyrians also possessed curious litters. "Two eunuchs," says Stratton, "are shown carrying a sort of arm-chair on their shoulders, elegant in design, supplied with wheels, to be drawn by hand should the king have occasion to visit mountainous regions inaccessible for chariots."

sometimes surrounded by a metal tyre. While only two horses were attached to the yokes, in the older monuments a third horse is generally to be seen, which was probably used as a reserve. The later chariots are square in front, not rounded; the car itself is larger and higher; the cases for the weapons are placed in front, not at the side; and only two horses are used. The harness differs somewhat from the Egyptian. A broad collar passes round the neck, from which hangs a breast ornament, the whole being secured by a triple strap under the belly of the horse. As in Egypt there are no traces visible; two driving-reins are attached to each horse, but the bearing-rein seems to be unknown. In addition to the warrior and the charioteer, we often see a third man who bears a shield; and a fourth occupant of the chariot sometimes appears.

"The Hittite chariots, as represented on Egyptian monuments, regularly contain three warriors. In construction they are plainer and more solid than the Egyptian, and the sides are not open. The chariots on Persian sculptures closely resemble the Assyrian."

There is still preserved in the Archæological Museum at Florence an Egyptian chariot, a light, simple, two-wheeled affair with a single shaft and four spokes to the wheels. From the number of spokes it may be supposed that this particular chariot was not used in war. In New York, too, there is preserved the wheel of an Egyptian chariot found at Dashour. The particulars of this bear out Mr. White's description. The wheel itself is three feet high, with a long axle arm, six spokes, tapering towards the felloe, and a double rim. "The six inner felloes do not meet as in modern wheels," says Thrupp,[8] "but are spliced one over the other, with an overlap of three inches."

Artificial roads seem to have existed at an early period in Palestine, but the country was hardly suitable for vehicles, and one first hears of waggons in the flatter wastes of Egypt and the level plains of Philistia. Agricultural carts these were, though no doubt early used for passenger traffic. Some of these carts were most probably covered, though no coverings seem to have been fixed to the chariots. The Assyrians, however, occasionally took into their private chariots an attendant, who was provided with a covering shaped somewhat like a modern umbrella. This covering was held over the owner's head, and was sometimes provided with a curtain which hung down at the back.

Details of the private carriages in use during these Biblical times filter through the chronicles. In Syria the merchants despatched by Solomon to buy chariots had to pay 600 shekels each for them. Solomon in his quest for luxury seems to have been the first man to build a more elaborate car than satisfied his contemporaries. One to be used on state occasions was built of cedar wood and had "pillars of gold." Probably it was some form of litter. The number of private cars was increasing enormously in all these Eastern cities. The prophet Nahum in lamenting the future woes of Nineveh speaks of "the noise of the whip, and the noise of the rattling of the wheels, and of the prancing horses, and of the jumping chariots,"

[8] *The History of Coaches.* G. A. Thrupp. London, 1877.

which will no longer bear witness to the city's prosperity. The absence of wide roads, however, militated against great changes of form in the carriages, which maintained their simple shape until many centuries later.

The war-chariot (ἅρμα or δίφρος) of the early Greeks was curved in front, and loftier than that of the Egyptians. The entrance was at the back. It was never covered, but frequently bore a curious basket-like arrangement, the πείρινς, upon or in which two people could sit. The ἄντυξ, or rim, in most cases ran round the three sides of the body, but occasionally there was only a curved barrier in front. The body itself was often strengthened by a trellis-work of strips of light wood or metal. The barrier was of varying height; in some chariots it did not reach above the driver's knee; in others it came up to his waist, but in war-chariots never higher than that. The axle was of oak, ash, elm, or even of iron, and precious metals, according to the legend, were used for the chariots of the gods. So of Juno's car we read:—

> "The whirling wheels are to the chariot hung.
> On the bright axle turns the bidden wheel
> Of sounding brass: the polish'd axle steel.
> Eight brazen spokes in radiant order flame;
> The circles gold, of uncorrupted frame,
> Such as the heavens produce; and round the gold
> Two brazen rings of work divine were roll'd.
> The bossy naves of solid silver shone;
> Braces of gold suspend the moving throne."

The last line suggests an innovation which was certainly not followed for some considerable time.

The chariot in general was about seven feet long, and could be lifted by a strong man like Diomed. Indeed, it could be driven over the bodies of dead warriors. The pole sloped sharply upwards, and sometimes ended in the head of a bird or animal. It emerged either from the floor of the car or from the axle. Towards its end the yoke for the horses was fastened about a pin fixed into it. Though the Lydians used chariots with two or even three poles, the Greeks never had more than one; and as with the Egyptians, there were no traces. If the pole broke, the horses must have dashed away with part of it, leaving the chariot at a standstill. Occasionally, too, a third horse was used, upon which sat a postilion.

At a later period several Grecian carriages were in common use, though not in warfare. Representations of such cars are to be found on the Elgin Marbles. And, as was the case a dozen or more centuries afterwards, the carriage became the outward sign of luxury. It invariably appeared in the state processions, and was made the receptacle for the most gorgeous ornamentation. Gold, ebony, copper, ivory, and white lead were all used for this purpose, while the interiors of the cars were made comfortable with soft cushions and fine tapestries. They appeared, too, in great numbers at the famous chariot races, at which four or more horses were driven abreast. Often the same man was rich enough to possess more than one carriage. So we read of Xerxes changing from his ἅρμα to his ἁρμάμαξα, or state-carriage, at the end of a march. Besides these, there were also the ἀπήνη, a kind of family sociable, the ἅμαξα, a waggon, the κάναθρον, and the φορεῖον, or litter.

The ἁρμάμαξα was a large four-wheeled waggon, enclosed by curtains and provided with a καμάρα or roof. Four or more horses were required to draw it. It was so large that a person could lie in it at full length, and, indeed, on many occasions it acted the part of a hearse. By far the most extraordinary hearse ever built was a ἁρμάμαξα used to convey the body of Alexander the Great—himself the possessor of numerous carriages—from Babylon to Alexandria.

> "It was prepared," says Thrupp, "during two years, and was designed by the celebrated architect and engineer Hieronymus. It was 18 feet long and 12 feet wide, on four massive wheels, and drawn by sixty-four mules, eight abreast. The car was composed of a platform with a lofty roof supported by eighteen columns, and was profusely adorned with drapery and gold and jewels; round the edge of the roof was a row of golden bells; in the centre was a throne, and before it the coffin; around were placed the weapons of war and the arms that Alexander had used."

The ἁρμάμαξα was also largely used by the ladies of Greece, who when they drove forth were careful to see that the curtains completely enclosed them. The ἅμαξα, also a four-wheeled waggon, was probably similar to the ἁρμάμαξα, though built upon a less imposing scale. The ἀπήνη was a still lighter carriage. It is described by Herodotus, and seems to have been a covered vehicle surrounded by silken curtains which could be pulled back when required. Its interior was generally furnished with cushions of goat leather. Two wheels were more frequent, but four were sometimes found. It was said that Timoleon, an old blind man, drove upon one occasion into the senate house and delivered a speech from his ἀπήνη. In some cases a two-wheeled carriage of this kind was not furnished with curtains, but enclosed in an oval-shaped covering of basket-work. Hesiod objected to such a conveyance because of its inability to keep out the dust. Little is known of the κάναθρον, but it was a Laconian car made of wood, with an arched, plaited covering, used chiefly by women. Doubtless it was little different from the ἀπήνη.

Coming to the Romans, we find a far greater variety of vehicles, though the descriptions that have come down are meagre and not particularly distinctive. That the Romans early realised the enormous importance, both military and otherwise, of carriages, is shown by their amazing roads. Such roads had never before been constructed. They were, says Gibbon, "accurately divided by milestones, and ran in a direct line from one city to another, with very little respect for the obstacles, either of nature or private property. Mountains were perforated, and bold arches thrown over the broadest and most rapid streams. The middle part of the road was raised into a terrace, which commanded the adjacent country, consisted of several layers of sand, gravel, and cement, and was paved with large stones, or, in some places near the capital, with granite." Probably the most famous of these roads was the Appian Way, connecting Rome with Capua. It was wide enough, according to Procopius, who marched along it in the sixth century, for two chariots to pass one another without inconvenience or delay, a matter certainly not possible, for instance, in most of the Eastern cities at that time. And so, with the finest engineers the world had seen linking up various cities, cross-country travelling in a carriage, from being well-nigh impossible, became comparatively easy. Gibbon mentions in this connection the surprising feat of one Cæsarius, who journeyed from Antioch to Constantinople, a distance of 665 miles, in six days.

The Roman war-chariot, or *currus*, was practically the same as the Greek ἅρμα, though certain modifications were introduced. More than two horses were driven, and from their number came several words, such as *sejugis, octojugis,* and *decemjugis,* which sufficiently explain themselves. It appears, moreover, that the *currus* was occasionally driven by four horses without either pole or yoke, and it has been suggested that in such a case the driver probably stopped the car by bearing all his weight on to the back of the body, so that its floor would touch the ground, thus forming a primitive brake. Besides the *currus*, and even before their marvellous roads had been laid down, the Romans possessed other cars. The earliest of these seems to have been a long, covered, four-wheeled waggon, called *arcera*, which was mainly used to carry infirm or very old people. In this the driver sat on a seat in front of the body, and drove two horses abreast. Though the most ancient of the Roman carriages, the *arcera*, as seen on monuments, has a very modern appearance. In more luxurious times the *lectica*, a large litter, seems to have led to its gradual extinction.

The *essedum*, at one time very popular in Italy, was brought in the first place to Rome by Julius Cæsar. It was the war-chariot of the Britons, and was entirely unlike the Roman or Egyptian cars. The wheels were much larger, the entrance was in front and not at the back, there was a seat, and the pole, instead of running up to the horses' necks, remained horizontal, and was so wide that the driver could step along it. The British charioteers could drive their cars at a very great rate, and were exceedingly agile on the flat pole, from the extremity of which they threw their missiles. The cars were purposely made as noisy as possible to strike dismay into the enemy's lines. At times the wheels were furnished with scythes, which projected from the axle-tree ends, and helped to maim those unfortunate enough to be run down.[9] Cicero, hearing good opinions of it, besought a friend to bring him a good pattern from Britain, and took occasion to add that the chariot was the only pleasing thing which that benighted country produced. The *essedum* speedily became popular in Rome, though not as an engine of war. Decorated and constructed of fine materials, it was the fashionable pleasure carriage. Curiously enough, however, the seat which had been so conspicuous a feature of the chariot in its native place was not used in Rome. The owner drove the *essedum* himself, and yoked two horses to the pole. There was some opposition to its use on the grounds of undue luxury, and a tribune who rode abroad in one was on that account considered effeminate. Seneca put the *esseda deaurata* amongst things *quæ matronarum usibus necessaria sint.* Emperors and generals used them as travelling carriages, and they were to be hired at regular posting-stations. A somewhat similar carriage, the *covinus*, was also in use in various countries at this date. This was covered in except

---

[9] See pp. 15-16.

in front; like the *essedum*, it had no seat for the driver, and in times of war it seems to have had scythes attached to the axle in the British fashion. Little, however, is known of it, and it may be dismissed here with a mere mention of its existence.

Cisium The Primitive Gig

(From a Roman Inscription)

Agrippina's Carpentum

(From a Roman Coin)

The *essedum* is of particular importance insomuch as it may be considered to be the prototype of all the vehicles of the curricle or gig type. The first of these in use amongst the Romans was the *cisium*, whose form is well shown on a monumental column near Treves. It was surprisingly like the ordinary gig of modern times. The body at first was fixed to the frames, but afterwards seems to have been suspended by rough traces or straps. The entrance was in front, there was a seat for two, and underneath this a large box or case. Mules were generally used to draw it, one, a pair, or, according to Ausonius, three—in which case a postilion sat on the third horse. They were built primarily for speed, and were in common use throughout Italy and Gaul, though the ladies, unwilling to be seen in an uncovered carriage, drove in other conveyances. The *cisium* on the whole must have been comfortable and light. Seneca admits that you could write a letter easily while driving in one. And in due course the new carriage became so popular that it could be hired, and the *cisiarii*, or hackney coachmen, could be penalised for careless driving. Indeed, so very modern were the Roman ideas upon the question of travel, that there were certain places at which the *cisium* was always to be found—a kind of primitive cab-rank.

Coming to the larger waggons and carriages, there were the *sarracum*, the *plaustrum*, the *carpentum*, the *pilentum*, the *benna*, the *reda*, the *carruca*, the *pegma*—a huge wheeled apparatus used for raising great weights, particularly in theatrical displays—and a mule-drawn litter, the *basterna*. Of these the *sarracum* was a common cart used by the country folk for conveying produce. It had either two or four wheels, and was occasionally used by passengers, though, as Cicero observed, as a conveyance the *sarracum* was very vulgar. It was not confined to Italy, but was common enough amongst those barbaric tribes against whom Rome was so often victorious. It was in *sarraca*, moreover, that the bodies were removed from Rome in times of plague. Rather lighter than this carriage, though heavy enough to our modern ideas, was the *plaustrum*,[10] an ancient two or four-wheeled waggon of rude construction. This was, in its primitive form, just a bare platform with a large pole projecting from the axle; there were no supporting ribs at all, and the load was simply placed on the platform. Upright boards, or openwork rails, however, were used to make sides, and at a later period a large basket was fastened on to the platform by stout thongs. The wheels of the *plaustrum* were ordinarily solid, of a kind called *tympana*, or drums, and were nearly a foot thick. Such a cart was but a slow vehicle, and could turn only with great difficulty. It was drawn by oxen or mules, and like the *sarracum* was also used to carry passengers.[11]

---

[10] Stratton treats of these Roman carriages and carts in considerable detail, and mentions in addition to the *plostellum*, or small *plaustrum*, the *carrus, monarchus*, and *birotum*. Of these the *carrus*, or cart, differed from the *plaustrum* in the following particulars: "The box or form could not be removed, as in the former case, but was fastened upon the axle-tree; it lacked the broad flooring of planks or boards, which served as a receptacle for certain commodities when the sides were removed; the wheels were higher [and] ... spoked, not solid like the *tympana*." The *carrus clabularius*, or stave-waggon, could be lengthened or shortened as required. The *monarchus* was a very light two-wheeled vehicle something like the *cisium*. The *birotum* was also a small two-wheeled vehicle, with a leather-covered seat, used in the time of Constantine, an "early post-chaise," as Stratton puts it.

[11] The carts of north Italy in the eighteenth century had remained practically un-

Pilentum

The State Carriage of the Romans

The *carpentum*, though two-wheeled, bore resemblance to the Greek ἁρμάμαξα. It had an arched covering. It was in use during very early times at Rome, though only distinguished citizens were privileged to ride in it. The *currus arcuatus*, given by Numa to the Flamines, was no doubt a form of *carpentum*, which was also the travelling carriage of the elder Tarquin. It seems to have been evolved from the *plaustrum*, being originally little more than a covered cart; but in the days of the Empire it became most luxurious, and was not only furnished with curtains of the

---

changed. Edward Wright, who visited Italy in 1719, thus describes them: "The carriages in Lombardy, and indeed throughout all Italy, are for the most part drawn with oxen; which are of a whitish colour: they have very low wheels. Some I saw without spokes, solid like mill-stones; such as I have seen describ'd in some antique basso-relievos and Mosaicks. The pole they draw by is sloped upwards towards the end; which is rais'd considerably above their heads; from whence a chain, or rope, is let down and fasten'd to their horns; which keeps up their heads, and serves to back the carriage. In some parts they use no yokes, but draw all by the horn, by a sort of a brace brought about the roots of them: the backs of the oxen are generally cover'd with a cloth. In the kingdom of Naples, and some other parts, they use buffaloes in their carriages, &c. These do somewhat resemble oxen: but are most sour, ill-looking animals, and very vicious; for the better management of them they generally put rings in their noses."

richest silk, but seems to have had solid panellings and sculptures attached to the body. Agrippina's *carpentum*, for instance, had fine paintings on its panels, and its roof was supported by figures at the four corners. Like the ἁρμάμαξα, it was also used as a hearse. Two mules were required to draw it. The *pilentum* was a carriage of a more official character. It may be called the state coach of the Romans—a four-wheeled becushioned car with a roof supported by pillars, but, unlike the *carpentum*, open at the sides. It was always considered to be the most comfortable of the Roman carriages, and may indeed have been hung upon "swing-poles" between the wheels. The social difference between the *pilentum* and the *carpentum* may be deduced from one of the many carriage laws passed by the Senate. The Roman matrons were allowed to drive in the *carpentum* on all occasions, but might use the *pilentum* only at the games or public festivals. Such "sumptuary laws" were constantly being passed, and a special vote was even required to enable the mother of Nero to drive in her carriage in the city itself. It was not until the fourth century A.D. that all such restrictions were banished.

Benna

Pliny mentions another carriage of imperial Rome—the *carruca*, which had four wheels and was used equally in the city and for long journeys. Nero travelled with great numbers of them—on one occasion with no less than three thousand. In Rome itself the fashionable citizen drove forth in a *carruca* that was covered with plates of bronze, silver, or even gold. Enormous sums were spent upon their decoration. Painters, sculptors, and embroiderers were employed. Martial speaks of an *aurea carruca* costing as much as a large farm. The *carruca*, indeed, may be said to correspond with the phaeton, which was so fashionable in England towards the end of the eighteenth century. As with the phaeton, so with the *carruca*—the higher it was built the better pleased was its owner. Various kinds of *carruca* existed. The *carrucæ argentatæ* were those granted by Alexander Severus to the senators. There is also mention of a *carruca domestoria*. Unfortunately, however, no contemporary representation of a carriage can definitely be said to be a *carruca*. Little enough, moreover, is known of the two other waggons, the *reda* and the *benna*. The *reda* was a large four-wheeled waggon used mainly to convey agricultural produce. It seems to have been brought into Italy from Wallachia. The *benna* was a cart whose body was formed entirely of basket-work. There is a drawing of it on the column of Antoninus at Rome. A similar vehicle persists to this day in Italy, South Germany, and Belgium, and bears a similar name.

Under the Empire, then, carriage-building flourished, particularly after Alexander Severus had put an end to all the older restrictions. Various forms of carriages were to be seen on the roads, and there was, as I have hinted, even an attempt at a spring. One of the carriages of this period is definitely described as "borne on long poles, fixed to the axles." "Now a certain amount of spring," says Thrupp, "can be obtained from the centre of a long, light pole. The Neapolitan Calesse, the Norwegian Carriole, and the Yarmouth Cart were all made with a view to obtaining ease by suspension on poles between bearings placed far apart. In these the seat is placed midway between the two wheels and the horse, on very long shafts, which are there made into wooden springs." And in the old Roman carriages, he goes on to say, "the weight was carried between the front and hind axles, on long poles or wooden springs. The undercarriage of the later four-wheeled vehicles used by the Romans was, in all probability, the same as is in use at the present day, both in this country and on the Continent, and indeed in America, for the under-carriages of agricultural waggons." Even with such splendid roads as the Romans possessed, however, the streets of their towns do not seem to have been very wide, and this must be one of the reasons for the early appearance of another kind of conveyance, the litter, which, during the dark ages, was practically the only carriage to be used.

These litters came from the East. The Babylonians in particular preferred to be carried about in a chair or couch rather than to be jolted in a carriage. Ericthonius, a lame man, is supposed to have introduced them into Athens, where they were known as φορεῖα or σκιμπόδια. Speedily they became popular, especially with the women. Magnificently decorated, the φορεῖον was constantly carried along the narrow streets, and on being brought over to Rome proved no less agreeable to the Romans. The *lectica*, or, as it was called at a later period, the *sella*, may in the first instance have been used to carry the sick, but in a short time became a common form of conveyance. This palanquin had an arched roof of leather stretched over four posts. The sides were covered by curtains, though at a later period it would seem that crude windows of talc were used. The interior was furnished with pillows, and when standing the litter rested upon four feet. Two slaves bore it by means of long poles loosely attached. In Martial's time these *lecticarii* wore red liveries, and were sometimes preceded by a third slave to make way. Julius Cæsar restricted their numbers, and in the reign of Claudius permission to use them was granted only as a particular mark of the royal favour. Several varieties of litter appeared. The *sella portatoria* or *gestatoria* was a small sedan chair. Some, however, were constructed to hold two. The *cathedra*, which was probably identical with the *sella muliebris* mentioned by Suetonius, was mostly used by women. The *basterna* was a much larger litter, also used by women under the Empire, which was carried by two mules. In this carriage the sides might be opened or closed, and the whole body was frequently gilded.

A few other primitive carriages here call for mention. The Dacians, who inhabited parts of what is now Hungary, used square vehicles with four wheels, in which the six spokes widened towards the rims. The Scythians used a peculiar two-wheeled cart consisting of a platform on which was placed a conical covering, resembling in shape a beehive, and made of a basket-work of hazelwood, over which were stretched the skins of beasts or a thatching of reeds. When camping out these people would lift this covering bodily from the cart and use it as a tent. Much the same custom was followed by the wandering Tartars. "Their huts or tents," says Marco Polo, "are formed of rods covered with felt, and being exactly round and nicely put together, they can gather them into one bundle, and make them up as packages, which they carry along with them in their migrations, upon a sort of car with four wheels." "Besides these cars," he continues, "they have a superior kind of vehicle upon two wheels, covered likewise with black felt, and so effectually as to protect those within it from wet during a whole day of rain. These are drawn by oxen and camels, and serve to convey their wives and children, their utensils, and such provisions as they require." The same traveller described the carriages of Southern China. Speaking of Kin-sai, then the capital, he says, "The main street of the city ... is paved with stone and brick to the width of ten paces on each side, the intermediate part being filled up with small gravel, and provided with arched drains for carrying off the rain-water that falls into the neighbouring canals, so that it remains always dry. On this gravel it is that the carriages are continually passing and re-passing. They are of a long shape, covered at top, having curtains and cushions of silk, and are capable of holding six persons. Both men and women who feel disposed to take their pleasure are in the daily practice of hiring them for that purpose, and accordingly at every hour you may see vast numbers of them driven along the middle part of the street." To this day such carriages as are here described can be had for hire in China, though in general they are of a smaller size. In some respects they resembled what is called in this country a tilted cart.

The Persians used large chariots in which was built a kind of turret from whose interior the warriors could at once throw their spears and obtain protection. One, taken from an ancient coin, is thus described by Sir Robert Ker Porter in his *Travels in Georgia, Persia, and Ancient Babylon* (1821):—

> " ... a large chariot, which is drawn by a magnificent pair of horses; one of the men, in ampler garments than his compeers, and bareheaded, holds the bridle of the horses ... [which] are without trappings, but the details of their bits and the manner of reining them are executed with the utmost care. The pole of the car is seen passing behind the horses, projecting from the centre of the carriage, which is in a cylindrical shape, elevated rather above the

line of the animals' heads. The wheel of the car is extremely light and tastefully put together."

Here, too, it is to be noticed that the driver is shown with his arms over the backs of the animals. In another chariot, which most probably was Persian, the body seems to be made of a "light wood, as of interlaced canes. Similar chariots are seen in the Assyrian bas-reliefs and others, somewhat resembling this, on Etruscan and Grecian painted vases. A chariot thus constituted must have been of extreme rapidity and of scarcely any weight."[12]

The Persians also had an idol-car, which was a kind of moving platform, and their chariots were at one period armed with scythes. These scythes, generally considered to be the invention of Cyrus, do not seem to have hung from the axle-ends, as was the case in Britain, but from the body itself, "in order," thinks Ginzrot, who wrote on these early carriages, "to allow the wheels to turn unobstructed. In this way," he says, "the scythes had a firm hold, and could inflict more damage than if they had been applied to the wheels or felloes and revolved with them. Nearly all writers treating on this subject are of this opinion, and Curtius says: *Alias deinde falces summis rotarum orbibus hærebant* [thence curving downwards]. The scythes could easily have been attached to the body ... and, notwithstanding, it might be said they extended over the felloe, for Curtius said, not that the scythes revolved with the wheels, but *hærebant*."[13]

Early Indian carriages were probably not very different from some of those now in use amongst the natives. The common *gharry* is certainly built after a primitive model. In this there are two wheels, "a high axle-tree bed, and a long platform, frequently made of two bamboos, which join in front and form the pole, to which two oxen are yoked." In Arabia there was the *araba*, a primitive latticed carriage for women, which possessed "wing-guards"—pieces of wood shaped to the top of the wheels and projecting over them—a feature also to be found in the early Persian cars.

Taking these early carriages as a whole one may be inclined to feel surprise at the varieties displayed, yet there were not after all very great differences between them. They were two-or four-wheeled contrivances with a long pole in front, and it is only in mere size and decoration that discrimination can properly be made. "The Egyptians," says Thrupp, "with all their learning and skill, appear to have made no change during the centuries of experience; as at the beginning, so at the end, the kings stand by the side of their charioteers, or hold the reins themselves. The Persians and Hindoos introduced luxurious improvements, and in lofty vehicles elevated the nobles above the heads of the people, and secluded their women

---

[12] *The World on Wheels.*

[13] On the other hand, the scythes used by other nations may well have been on the wheels. Livy describes those used by Antiochus (*currus falcatus*): "Round the pole were sharp-pointed spears which extended from the yoke of the two outside horses about fifteen feet; with these they pierced everything in their way. On the end of the yoke were two scythes, one being placed horizontally, the other towards the ground. The first cut everything from the sides, the others catching those prostrate on the ground or trying to crawl under. The long spears (*cuspides*) were not on the yoke, as some say."

in curtained carriages. The Greeks introduced no new vehicles, but perfected so successfully the useful waggon, that their model is still seen throughout Europe, without change of principle or structure. The Romans, on the other hand, in their career of conquest, gathered from every nation what was good, and, wherever possible, improved upon it." After the fall of the Roman Empire, however, there was little further progress for several centuries. In the general retrogression, which, rightly or wrongly, one associates with those dark ages, the wheeled carriage, in common with a multitude of other adjuncts to civilisation, was to suffer.

CHAPTER THE SECOND

THE AGE OF LITTERS

"There is a litter; lay him in 't and
drive toward Dover, friend!"

King Lear.

As roadmakers, the Romans, if they can be said to have had successors at all, were succeeded by the monks. On the assumption that travellers were unfortunate people, as indeed they were, needing help, religious Orders were founded whose chief work was that of building bridges and repairing the roads. Other Orders likewise performed such tasks, though possibly for more selfish reasons, being as they were large owners of cattle, and immersed as much in agricultural as in theological occupations. So in many parts of Europe the *Pontife* Brothers, or bridge-makers, were to be found. There were also Gilds formed to repair the roads, such as the Gild of the Holy Cross in Birmingham, founded in the reign of Richard II, which "mainteigned ... and kept in good reparaciouns the greate stone bridges, and divers foule and dangerous high wayes, the charge whereof the towne of hitsellfe ys not hable to mainteigne." In *Piers the Plowman*, too, the rich merchants are exhorted to repair the "wikked wayes" and see that the "brygges to-broke by the heye weyes" may be mended "in som manere wise." The maintenance of the roads in England, says M. Jusserand,[14] "greatly depended upon arbitrary chance, upon opportunity, or on the goodwill or the devotion of those to whom the adjoining land belonged. In the case of the roads, as of bridges, we find petitions of private persons who pray that a tax be levied upon those who pass along, towards the repair of the road." So in 1289, Walter Godelak of Walingford is praying for "the establishment of a custom to be collected from every cart of merchandize traversing the road between Jowemarsh and Newenham, on account of the depth, and for the repair, of the said way." Unfortunately for him—and doubtless he was no exception to the rule—the reply came: "The King will do nothing therein."

Indeed the roads were in a truly abominable condition. As often as not, deep ruts marred what surface there had ever been, and here and there brooks and pools rendered easy passage an impossibility. There is a patent of Edward III (Nov. 20, 1353) which ordered "the paving of the high road, *alta via*, running from Temple Bar"—then the western limit of London—"to Westminster." "This road,"

---

[14] *English Wayfaring Life in the Middle Ages.* J. J. Jusserand. London, 1888.

says M. Jusserand, "had been paved, but the King explains that it is 'so full of holes and bogs ... and that the pavement is so damaged and broken' that the traffic has become very dangerous for men and carriages. In consequence, he orders each proprietor on both sides of the road to remake, at his own expense, a footway of seven feet up to the ditch, *usque canellum*," and see to it that the middle of the road is well paved. In France matters were just as bad. "Outside the town of Paris," runs one fourteenth-century ordinance, "in several parts of the suburbs ... there are many notable and ancient high-roads, bridges, lanes, and roads, which are much injured, damaged or decayed and otherwise hindered by ravines of water and great stones, by hedges, brambles, and many other trees which have grown there, and by many other hindrances which have happened there, because they have not been maintained and provided for in time past; and they are in such a bad state that they cannot be securely traversed on foot or horseback, nor by vehicles, without great perils and inconveniences; and some of them are abandoned at all parts because men cannot resort there." Wherefore it was proposed that the inhabitants should be compelled, by force if necessary, to attend to the matter.

While, however, the wretched state into which the roads were being allowed to fall had a great deal to do with the almost total, though indeed temporary, extinction of the wheeled pleasure carriage in western Europe, there is another fact which must be taken into consideration in any endeavour to account for it. As will appear in a little, the renaissance of carriage-building in the sixteenth century was for a time retarded in various places by a widespread feeling of distrust against anything that could be thought to lead to an accusation of effeminacy. Laws were passed—as was the case, for instance, in 1294, under Philip the Fair of France—forbidding people to ride in coaches, and sharp comparisons were drawn by the satirists between the hardy horsemen of old and the modern comfort-loving individuals who lolled, or were supposed to loll—though how they could have done so in those springless monstrosities is past comprehension—in their gaudily decorated carriages. I would not insist upon the point, but it may be that in the reaction against such undue luxuries as had helped to bring ruin to the Roman Empire, carriages for that reason became unpopular. From which, of course, it would follow that the disappearance of the carriage led, in part at any rate, to the neglect of the roads, and such new roads as were made would be laid down primarily for the convenience only of the horsemen. The same thing applied also to the litters, though their popularity naturally followed merely upon the state of the roads.

Before attempting to deal with these litters, it will be well to see what is known—it is not very much—of such wheeled carriages as there were at this time, and at the outset it is necessary to bear in mind that the old chroniclers used the word *carriage* in anything but its modern significance. To them a carriage was no more than an agricultural or baggage cart. Time and again you have accounts of this or that great man making his way, peaceably or otherwise, through some country, accompanied by numbers of carriages. These were simply his luggage carts, and although, as in earlier times, the cart, gaily ornamented, could very easily be converted into a pleasure carriage, it is important to remember the real meaning of the word. Such carts, in point of fact, were extremely common. In England they were generally square boxes made of planks borne on two wheels. Others, of a lighter pattern, were built of "slatts latticed with a willow trellis." Their chief peculiarity was to be found in their wheels, which were furnished with extraordinarily large nails with prominent heads. Contemporary manuscripts give rough pictures of such carts. One of these is shown drawn by three dogs. One man squats inside, a second helps to push it from behind. A most interesting illustration in the Louterell Psalter—a fourteenth-century manuscript—shows a reaper's cart going uphill. Here the two huge, six-spoked wheels with their projecting nails are clearly shown. The platform of the cart is strengthened by upright stakes with a cross-rail connecting them at the sides. The driver, standing over the wheels on the poles, is holding a long whip which is flicking the leader of three horses. Three other men are helping at the rear, and the stacks of wheat are held in position by ropes.

The earliest Anglo-Saxon carriage of which there is record belongs to the twelfth century. Strutt refers to a drawing in one of the Cottonian manuscripts, which represents a peculiar four-wheeled contrivance with two upright poles rising from the axle-trees, from which poles is slung a hammock. Such a chariot or *chaer* was apparently used by the more distinguished Anglo-Saxons when setting out upon long journeys. The drawing shows the figure of Joseph on his way to meet Jacob in Egypt, but is no doubt a correct representation of a travelling carriage in the artist's lifetime. This hammock is interesting as being a primitive form of suspension, which may or may not have led to the later experiments in that direction.

Fourteenth Century English Carriage

(From the Louterell Psalter)

Fourteenth Century Reaper's Cart

(From the Louterell Psalter)

A most luxurious English carriage of the fourteenth century is shown in the Louterell Psalter. This was obviously evolved from a four-wheeled waggon. Five horses, harnessed at length, drew it, a postilion with a short whip riding on the second, and another with a long whip on the wheeler. The tunnel-like body was highly ornamented, and its front decorated with carved birds and men's heads. The frame of the body was continued in front as two poles, and underneath, hanging by a ring and looking rather ludicrous, is shown a small trunk. Women only appear in this carriage, the men riding behind it.

"Nothing," remarks M. Jusserand, "gives a better idea of the encumbering, awkward luxury which formed the splendour of civil life during this century than the structure of these heavy machines. The best had four wheels; three or four horses drew them, harnessed in a row, the postilion being mounted on one, armed with a short-handled whip of many thongs; solid beams rested on the axles, and above this framework rose an archway rounded like a tunnel; as a whole, ungraceful enough. But the details," he goes on to say, speaking of the carriage shown in the Louterell Psalter, "were extremely elegant, the wheels were carved and their spokes expanded near the hoop into ribs forming pointed arches; the beams were painted and gilt, the inside was hung with those dazzling tapestries, the glory of the age; the seats were furnished with embroidered cushions; a lady might stretch out there, half sitting, half lying; pillows were disposed in the corners as if to invite sleep, square windows pierced the sides and were hung with curtains. Thus travelled," he continues with a touch of picturesqueness, "the noble lady, slim in form, tightly clad in a dress which outlined every curve of the body, her long, slender hands caressing the favourite dog or bird. The knight, equally tightened in his *cote-hardie*, regarded her with a complacent eye, and, if he knew good manners, opened his heart to his dreamy companion in long phrases like those in the romances. The broad forehead of the lady, who has perhaps coquettishly plucked off her eyebrows and stray hairs, a process about which satirists were indignant, brightens up at moments, and her smile is like a ray of sunshine. Meanwhile the axles groan, the horse-shoes—also heavily nailed—crunch the ground, the machine advances by fits and starts, descends into the hollows, bounds altogether at the ditches, and falls violently back with a dull noise."

Other gaily decorated carriages, surprisingly like our modern vans, though on two wheels, are shown in *Le Roman du Roy Meliadus*, another fourteenth-century manuscript preserved in the British Museum, but only the richest and most powerful of the nobles could afford to keep them.

"They were bequeathed," says M. Jusserand, "by will from one another, and the gift was valuable. On September 25, 1355, Elizabeth de Burgh, Lady Clare, wrote her last will and endowed her eldest daughter with 'her great carriage with the coverture, carpets, and cushions.' In the twentieth year of Richard II, Roger Rouland received £400 sterling for a carriage destined for Queen Isabella; and John le Charer, in the sixth [year] of Edward III, received £1000 for the carriage of Lady Eleanor—the King's sister."

These were fabulous sums, when it is remembered that an ox cost about thirteen shillings and a sheep but one shilling and five pence.

Now it may be that such a "great carriage" as is shown in the Louterell Psalter was identical with the *whirlicote* in which, according to Stowe, Richard II and his mother took refuge on the occasion of Wat Tyler's rebellion.

"Of old time," says this honest tailor, who himself witnessed the introduction of coaches into England, "coaches were not known in this island, but chariots or whirlicotes, then so called, and they only used of princes or great estates, such as had their footmen about them; and for example to note, I read that Richard II, being threatened by the rebels of Kent, rode from the Tower of London to the Mile's End, and with him his mother, because she was sick and weak, in a whirlicote, the Earl of Buckingham ... knights and Esquires attending on horseback. But in the next year (1381) the said King Richard took to wife Anne, daughter to the

King of Bohemia, that first brought hither the riding upon side saddles; and so was the riding in whirlicotes and chariots forsaken, except at coronations and such like spectacles."

From this it would appear that the *whirlicote* (which may, as Bridges Adams suggests, have been derived from "whirling" or moving "cot" or house) was identical with the *chariot* or *chaer*. Unfortunately the translators of Froissart, who mentions the incident of Richard's ride from the Tower, cannot agree upon the correct word to render the original *charette*. *Charette, chariette, chare, chaer* (Wicliffe), and *char* (Chaucer) all occur in the early chronicles, and there seems no means, if, indeed, there is any need, of differentiating between them. All were probably waggons modified for the conveyance of such passengers as could afford to pay highly for the privilege. One fact, however, suggests that there were at any rate two different kinds of carriages in England at this time, for we read that the body of Richard II was borne to its last resting-place "upon a chariette or sort of litter on wheels, such as is used by citizens' wives who are not able or not allowed to keep ordinary litters." With this in mind, it is difficult to agree with Sir Walter Gilbey when he says[15] that the *chare* was a horse litter, though it is fair to add that he acknowledges an opposite view.

The *charette* is obviously the French form of *caretta*, which was the carriage in which Beatrice, the wife of Charles of Anjou, entered Naples in 1267.[16] This vehicle is described as being covered both inside and without with sky-blue velvet powdered with golden lilies. Pope Gregory X entered Milan in 1273 in a similar carriage. The *caretta* was probably an open car "shaded simply by a canopy." In the next century, the *Anciennes Chroniques de Flandres*, a manuscript belonging to 1347, shows an illustration of Ermengarde, the wife of Salvard, Lord of Rousillon, travelling in a four-wheeled conveyance remarkably like the ordinary country waggon of to-day.

"The lady," says Sir Walter Gilbey, "is seated on the floorboards of a springless four-wheeled cart or waggon, covered in with a tilt that could be raised or drawn aside; the body of the vehicle is of carved wood and the outer edges of the wheels are painted grey to represent iron tyres. The conveyance is drawn by two horses driven by a postilion who bestrides that on the near [left] side. The traces are apparently of rope, and the outer trace of the postilion's horse is represented as passing under the saddle-girth, a length of leather (?) being let in for the purpose; the traces are attached to swingle-bars carried on the end of a cross-piece secured to the base of the pole where it meets the body.

---

[15] *Early Carriages and Roads*. Sir Walter Gilbey, Bart. London, 1903.

[16] This appears to have been similar to the *carroccio*, described by Stratton as a very heavy four-wheeled car, surmounted by a tall staff, painted a bright red. Stratton also mentions the *cochio*, which he describes as a thirteenth-century carriage having a covering of red matting, under which, in the fore-part of the body, the ladies were seated, the gentlemen occupying the rear end. Both these *words*, however, seem to belong to a much later date and may be translations of an earlier original.

"Carriages of some kind," he continues, "appear also to have been used by men of rank when travelling on the Continent. *The Expeditions to Prussia and the Holy Land of Henry, Earl of Derby, in* 1390 *and* 1392-3 (Camden Society's Publications, 1894) indicate that the Earl, afterwards King Henry IV of England, travelled on wheels at least part of the way through Austria.

"The accounts kept by his Treasurer during the journey contain several entries relative to carriages; thus on November 14, 1392, payment is made for the expenses of two equerries named Hethcote and Mansel, who were left for one night at St. Michael, between Leoban and Kniltefeld, with thirteen carriage horses. On the following day the route lay over such rugged and mountainous country that the carriage wheels were broken despite the liberal use of grease; and at last the narrowness of the way obliged the Earl to exchange his own carriage for two smaller ones better suited to the paths of the district.

"The Treasurer also records the sale of an old carriage at Friola for three florins. The exchange of the Earl's 'own carriage' is the significant entry: it seems very unlikely that a noble of his rank would have travelled so lightly that a single cart would contain his own luggage and that of his personal retinue; and it is also unlikely that he used one luggage cart of his own. The record points directly to the conclusion that the carriages were passenger vehicles used by the Earl himself."

It is to be noted that the carriage of the Lady Ermengarde was a Flemish vehicle. Flanders, indeed, seems to have shared with Hungary the honour of playing pioneer in carriage-building throughout the ages, and long after the general adoption of coaches in Europe, Flemish models, and also Flemish mares, were freely imported into the various countries.

Another carriage of this time is described in a pre-Chaucerian poem called *The Squyr of Low Degree,* in which the father of a Hungarian princess is made to say:—

"Tomorrow ye shall on hunting fare,
And ride my daughter in a *chare.*
It shall be covered with velvet red,
And cloths of fine gold all about your head;
With damask white, and azure blue,
Well diapered with lilies new;
Your pomelles shal be ended with gold,
Your chains enammelled many a fold."

The pomelles no doubt were "the handles to the rods affixed to the roof, and were for the purpose of holding on by, when deep ruts or obstacles in the road caused an unusual jerk in the vehicle." One notices that lilies were apparently a common form of decoration on these early carriages, but it is to be regretted that the accounts in general are so scanty.

We come to the litters.

Of these the commonest, both in England and on the Continent, seem to have been modifications of the Roman *basterna*. Generally they were covered with a sort of vault with various openings. Two horses, one at either end, carried them. The great majority held only one person. Thrupp describes them in some detail.

> "They were," he says, "long and narrow—long enough for a person to recline in—and no wider than could be carried between the poles which were placed on either side of the horses. They were about four to five feet long, and two feet six inches wide, with low sides and higher ends. The entrance was in the middle, on both sides, the doors being formed sometimes by a sliding panel and sometimes simply by a cross-bar. The steps were of leather or iron loops, the latter being hinged to turn up when the litter was placed on the ground. The upper part was formed by a few broad wooden hoops, united along the top by four or five slats, and over the whole a canopy was placed, which opened in the middle, at the sides, and ends, for air and light."

Isolated references to these horse-litters are scattered throughout the old chronicles, but afford meagre information. William of Malmesbury states that the body of William Rufus was placed on a *reda caballaria*, a horse-litter, the name of which suggests its origin. According to Matthew of Westminster, King John, during his illness in 1216, was removed from Swinstead Abbey to Newark in a similar vehicle, the *lectica equestre*. Generally, however, the horse-litter was reserved exclusively for women, men being unwilling to risk an accusation of effeminacy. So, in recording the death of Earl Ferrers in 1254, from injuries received in an accident to his conveyance, the historian is careful to explain that his Lordship suffered from the gout, which was why he happened to be in a litter at all.

As time passed, the litter rather than the wheeled carriage became the state vehicle. Froissart, writing of the second wife of Richard II, describes "la june Royne d'Angleterre" as travelling "en une litere moult riche qui etoit ordondée pour elle." Margaret, the daughter of Henry VII, journeyed to Scotland, it is true, on the back of a "faire palfrey," but she was followed by "one vary riche litere, borne by two faire coursers vary nobly drest; in wich litere the sayd queene was borne in the intryng of the good townes, or otherwise to her good playsher." But on the Continent new improvements were being made in wheeled carriages, and when in 1432 Henry VI wrote to the Archbishop of Canterbury and other high dignitaries of the Church, with regard to the widow of Henry of Navarre, he ordered them to place two *chares* at her disposal, rather than the litter to which one might have thought she would be entitled. Sir Walter Gilbey translates the word to mean a horse-litter, but Markland, in his paper on the *Early Use of Carriages in England* (*Archæologia*, Vol. XX), differentiates between the two, ascribing a more ceremonial use to the litter, and this seems to me to be nearer the truth. Both vehicles, for instance, are mentioned by Holinshed in his description of the coronation ceremony of Catherine of Aragon in 1509. The Queen herself rode in a litter of "white clothe of golde, not covered nor bailed, which was led by two palfreys clad in white damask doone to the ground, head and all, led by her footman. Over her was borne a canopie of cloth of gold, with four gilt staves, and four silver bells. For the bearing of which canopie were appointed sixteen knights, foure to beare it one space on foot, and other foure another space." But the Queen's ladies followed her in *chariots* decorated in red, and the same thing is true of Anne Boleyn, who in 1533 rode to her coronation in a litter, but was followed by four chariots, three decorated with red, and one with white. Such chariots probably resembled those to be described in the next chapter; the point to notice here is that they were being used now, and although the litters still continued until the time of Charles II—Mary de Medicis, the Queen-Mother of France, entered London in 1638 in a litter, though she had travelled from Harwich in a coach, and as late as 1680 "an accident happened to General Shippon, who came in a horse-litter wounded to London; when he paused by the brewhouse in St. John Street a mastiff attacked the horses, and he was tossed like a dog in a blanket"—the wheeled carriage once again became the vehicle of honour, and at the coronation of Mary in 1553 a chariot[17] and not a litter was used by the Queen. This had six horses, and was covered with a "cloth of tissue." Whatever its discomforts may have been, it cannot have been less dignified than the litter which it had, now for all time, supplanted.

---

[17] "The xxx day of September the Queen's Grace came from the Tower through London, riding in a charrett gorgeously beseen, unto Westminster." MS. Cotton. Vitellius, F.v.

CHAPTER THE THIRD

INTRODUCTION OF THE COACH (1450-1600)

"Go—call a Coach; and let a Coach be called:
Let him that calls the Coach, be called the Caller!
And in his calling, let him no thing call,
But Coach! COACH!! COACH!!!"

Chrononhotonthologos.

BOTH horse-litters and early wheeled carriages seem to have had some pretensions towards comfort. They afforded protection against the inclemency of the weather; there had been certain rude attempts at suspension, and the soft cushions helped to minimise the unpleasant joltings to which every carriage was liable. When, however, the renaissance of carriage-building occurred, people seem to have been but little more progressive than they had been centuries before. There were, as I have already hinted, still two factors which militated against a speedy adoption of such vehicles, more comfortable though they undoubtedly were, as now began to be made—the state of the roads, and the dislike of anything bordering upon the effeminate.

The roads had become no better. Even those most eager to welcome the new carriages must have been dismayed at the state of the country, not only in England, but in every European country. As one writer of the sixteenth century complains, the roads, "by reason of straitness and disrepair, breed a loathsome weariness to the passenger." Nor is this writer a solitary grumbler: there are numerous complaints. In 1537 Richard Bellasis, one of the monastery-wreckers, was unable to proceed with his work: "lead from the roofs," he reports, "cannot be conveyed away till next summer, for the ways in that countrie are so foule and deepe that no carriage [cart] can pass in winter." Indeed, no one seems to have looked after the roads with any care, either in the fifteenth or the sixteenth century. Yet there were, in this country, repeated bequests for their preservation. Henry Clifford, Earl of Cumberland, a sufferer himself, left one hundred marks to be bestowed on the highways in Craven, and the same sum on those of Westmorland. John Lyon, the founder of Harrow School, gave certain rents for the repair of the roads from Harrow and Edgware to London. This was in 1592, and Lyon's example was speedily followed by Sutton, the founder of the Charterhouse. There was, indeed, legislation of a kind, but in general the roads were in a terrible condition, and for a long time, so far as men were concerned, the saddle remained triumphant.

And for an even longer time continued that prejudice against carriages which led to the framing of actual prohibitive laws. Even women were occasionally forbidden the use of coaches, and there is the story of the luxurious duchess who in 1546 found great difficulty in obtaining from the Elector of Saxony permission to be driven in a covered carriage to the baths—such leave being granted only on the understanding that none of her attendants were to be allowed the same privilege. So, too, in 1564, Pope Pius IV was exhorting his cardinals and bishops to leave the new-fangled machines to women, and twenty-four years later Julius, Duke of Brunswick, found it necessary to issue an edict—it makes quaint reading now— ordering his "vassals, servants, and kinsmen, without distinction, young and old," who "have dared to give themselves up to indolence and to riding in coaches ... to take notice that when We order them to assemble, either altogether or in part, in Times of Turbulence, or to receive their Fiefs, or when on other occasions they visit Our Court, they shall not travel or appear in Coaches, but on their riding Horses." More stringent is the edict, preserved amongst the archives of the German county of Mark, in which the nobility was forbidden the use of coaches "under penalty of incurring the punishment of felony." So, also, we have the case of René de Laval, Lord of Bois-Dauphin, an extremely obese nobleman living in Paris, whose only excuse for possessing a coach was his inability to be set upon a horse, or to keep in that position if the horse chanced to move. This was in 1550. In England there was a similar feeling of opposition. In 1584 John Lyly, in his play *Alexander and Campaspe*, makes one of his characters complain of the new luxury. In the old days, he says, those who used to enter the battlefield on hard-trotting horses, now ride in coaches and think of nothing but the pleasures of the flesh. The once famous Bishop Hall speaks bitterly of the "sin-guilty" coach:—

> "Is't not a shame to see each homely groome
> Sit perched in an idle chariot roome
> That were not meete some pannel to bestride
> Sursingled to a galled hackney's hide?
> Nor can it nought our gallant's praises reap,
> Unless it be done in staring cheap
> In a sin-guilty coach, not closely pent,
> Jogging along the harder pavement."

Possibly the same idea is to be found in the framing of a Parliamentary Bill of 1601 "to restrain the excessive use of coaches," which, however, was thrown out. So again in 1623, the delightful though sadly biased water-poet, John Taylor, is lamenting the decadence of England, due, according to him, to the growing custom of driving in coaches.

> "For whereas," he says, "within our memories, our Nobility
> and Gentry would ride well mounted (and sometimes walke on
> foote) gallantly attended with three or four, score brave fellowes
> in blue coates, which was a glory to our Nation; and gave more
> content to the beholders, then [*sic*] forty of your Leather tumbrels:
> Then men preserv'd their bodies strong and able by walking, rid-

ing, and other manly exercises: Then saddlers was a good Trade, and the name of a Coach was Heathen Greek. Who ever saw (but upon extraordinary occasions)," he goes on to ask, "Sir *Philip Sidney*, Sir *Francis Drake*, Sir *John Norris*, Sir *William Winter*, Sir *Roger Williams*, or (whom I should have nam'd first) the famous Lord *Gray* and *Willoughby*, when the renowned *George* Earle of *Cumberland*, or *Robert* Earle of *Essex*? These sonnes of *Mars*, who in their time were the glorious Brooches of our Nation, and admirable terrour to our Enemies: these, I say, did make small use of Coaches, and there were two mayne reasons for it, the one was, that there were but few Coaches in most of their times: and the second is, they were deadly foes to all sloth and effeminacy."

To Taylor, indeed, and probably to every one of his fellow-watermen, a coach was always a "hell-cart" designed on purpose to put an end to his own most worthy calling. But less biased poets than outspoken Taylor gave tongue to an opposition which continued for nearly two centuries. Gay, for instance, looked on the vastly improved vehicle of his day as no more than an excuse for extravagant display:—

"O happy streets, to rumbling wheels unknown,
No carts, no coaches shake the floating town!
Thus was of old *Britannia's* city bless'd,
Ere pride and luxury her sons profess'd."

And again:—

"Now gaudy pride corrupts the lavish age,
And the streets flame with glaring equipage;
The tricking gamester insolently rides,
With *Loves* and *Graces* on his chariot's sides;
In saucy state the griping broker sits,
And laughs at honesty, and trudging wits."

Perhaps he is thinking of some personal inconvenience, rather than of mere unnecessary luxury, when he asks:—

"What walker shall his mean ambition fix
On the false lustre of a coach and six?"

And so late as 1770, the eccentric Lord Monboddo, who still maintained the superiority of a savage life, refused to "sit in a box drawn by brutes." It is, of course, easy to magnify such opposition to coaches as followed on the grounds of mere luxury and display, but in the earlier history of the coach, to which we are now come, it is a factor which must by no means be neglected. The coach, like every other novelty, had to fight its way, and if one is inclined to believe, after reading such accusations as there are of the earliest coaches with their magnificent adornments and numerous attendants, that the owners altogether deserved the reproaches of their more Spartan fellows, it may be well to recall Macaulay's words. In his sketch of the state of England in 1685, when coaches were still lavishly adorned, he says of them: "We attribute to magnificence what was really the effect of a very disagreeable necessity. People in the time of Charles the Second travelled with six horses, because with a smaller number there was great danger of sticking fast in the mire." And what is true of 1685 is certainly true of 1585.

Buckingham is supposed to have been the first man to use a coach and six in this country, though this is by no means certain. Of him a well-known story apropos of this question of undue luxury is told. "The stout old Earl of *Northumberland*," it runs, "when he got loose, hearing that the great Favourite *Buckingham* was drawn about with a Coach and six horses (which was wondered at then as a *novelty*, and imputed to him as a *mastring pride*) thought if *Buckingham* had six he might very well have eight in his Coach, with which he rode through the City of *London* to the *Bath*, to the vulgar talk and admiration.... Nor did this addition of two horses by *Buckingham* grow higher than a little *murmur*. For in the late Queen's time there were no coaches, and the first [had] but two Horses; the rest crept in by *Degrees* as men at first venture to *sea*."[18] Yet what may have been true of Buckingham, whose love of luxury was notorious, need not have been true of those other owners of coaches, who were constantly travelling about the country.

Finally there is the other side of the question to be remembered, and, as M. Ramde quaintly points out in his *History of Locomotion*, the very luxury which people so disliked had a beneficent effect; for "after the development of the use of carriages, and their frequent employment by the court and nobility, the liberty to throw everything out of the window became intolerable! Thus the carriage of luxury has been the cause of cleanliness in the streets."

Now it must be understood that the coach proper differs from all earlier vehicles in being not only a covered, but also a suspended carriage. The canopy has given place to the roof, a roof, that is to say, which forms part of the framing of the body; and the body itself is swung in some fashion, however primitive, from posts or other supports. Further, it seems reasonable to suppose, on the analogy of the *berlin* and the *landau*—two later carriages which took their names from the towns in which they were first made—that the first coaches were built in a small Hungarian town then called Kotzee. Yet it is to be observed that Spain, Italy, and France, in the persons of various enthusiasts, have claimed the invention—their claims being mainly based on such similarities as may be observed between the real coach and the earlier cars and charettes.[19] Bridges Adams, indeed, not to be outdone, hazards the suggestion that England might also be included in such a list by reason of her invention of the whirlicote, though he is obliged to admit that nobody knows *exactly* what a whirlicote was like. It is probably due to these patriotic

---

[18] *History of Great Britain.* Arthur Wilson. London, 1653.

[19] cf. Spenser, who uses three words which appear to be interchangeable.

> "Tho', up him taking in their tender hands
> They easily unto her charett beare;
> Her teme at her commandement quiet stands,
> Whiles they the corse unto her wagon reare.
> And strowe with flowers the lamentable beare;
> Then all the rest into their Coches climb."

gentlemen that several rather ludicrous suggestions have been made to explain the derivation of the word *coach*, which has a similar sound in nearly all European languages. Menange rashly suggests a corruption of the Latin *vehiculum*. Another writer puts forward the Greek verb ὀχέω, to carry. Wachten, a German, finds in *kutten*, to cover, a suitable explanation, and Lye produces the Flemish *koetsen*, to lie along. This last, perhaps, is the most reasonable suggestion of those unwilling to give the palm to Hungary, for not only were the Flemish vehicles well known before the introduction of the new carriage, but there is also some confusion, at any rate, in this country, between the two words *coach* and *couch*, both being found in the old account books. Even in the sixteenth century the word seems to have bothered people. There is an amusing reference to this point in an early seventeenth-century tract called *Coach and Sedan Pleasantly Disputing*, of which I shall have more to say in the next chapter.

> "Their first invention," says a character in this dialogue, "and use was in the Kingdome of *Hungarie*, about the time when *Frier George*, compelled the Queen and her young sonne the King, to seeke to *Soliman* the Turkish Emperour, for aid against the Frier, and some of the Nobilitie, to the utter ruine of that most rich and flourishing Kingdome, where they were first called *Kottcze*, and in the *Slavonian* tongue *Cottri*, not of *Coucher* the French to lie-downe, nor of *Cuchey*, the Cambridge Carrier, as some body made Master *Minshaw*, when hee (rather wee) perfected his Etymological dictionarie, whence we call them to this day *Coaches*."

It is also to be noted that the first English coaches, so called, were probably not suspended at all, but merely upholstered carts for reclining—in fact nothing more than the old chariots. In the second half of the sixteenth century, practically every pleasure carriage in England, though not on the Continent, was called a *coach* or a *carroche*. Consequently it is difficult to give a date for the importation of the first real coach into this country. Indeed, it is impossible to say with any degree of certainty precisely when carriages of the suspended type were first made. Such early accounts as exist are at once fragmentary and obscure, and the few illustrations little better than caricatures with a perspective reminiscent of that in Hogarth's famous example of false drawing. It can only be repeated that the hammock slung from the four posts of a waggon, such as we have seen existed amongst the Anglo-Saxons and possibly was also in use in parts of Europe, may have provided the idea of permanent suspension as a means to comfort, and that such scanty evidence as there is goes to prove that the carriages exported from Hungary towards the end of the fifteenth century seem to have been the first *coaches* to be built.

So early as 1457 there is mention of such a carriage, given by Ladislaus, King of Hungary, to the French King, Charles VII. The Parisians who saw it described it as "branlant et moulte riche." What this "trembling" carriage was like there is no means of discovering, but it certainly suggests an attempt at suspension, and may perhaps be taken for the earliest coach to be recorded by history. This obviously was Hungarian, and Hungary is again mentioned in the same connection by Stephanus Broderithus, who relates that in 1526, "when the archbishop received intelligence that the Turks had entered Hungary, not content with informing the King of this event, he speedily got into one of those light carriages which from the name of the place we call kotcze, and hastened to His Majesty." And apparently these light carriages were actually used for military purposes, Taylor avowing that "they carried soldiers on each side with cross-bowes," this being the best purpose to which he considered the coach had ever been put or was likely to be put in the future. All this is clear enough, but Beckmann, in his *History of Inventions*, mentions another circumstance which strengthens the evidence: "Siegmund, Baron de Herberstein, ambassador from Louis II, to the King of Hungary, says in his *Commentarie de rebus Moscoviticis*, where he occasionally mentions some travelling-stages in Hungary: 'The fourth stage for stopping to give the horses breath is six miles below Taurinum, in the village of Cotzi, from which both drivers and carriages take their name, and are generally called cotzi.'"[20]

Very probably these new Hungarian carriages were seen in most European countries before 1530. "At tournaments," says Bridges Adams, "they were made objects for display; they are spoken of as being gilded all over, and the hangings were of crimson satin. Electresses and duchesses were seldom without them; and there was as much rivalry in their days of public exhibition as there is now (1837) amongst the aspirants of fashion in their well-appointed equipages at a queen's drawing-room."

What did these early coaches look like? Shorn of their hangings, they must have resembled nothing so much as the hearse of to-day. The first illustrations show no signs of suspension, and portray what appear to be gaudily decorated waggons, and that in effect is what they were. The first coach makers of Hungary, like their predecessors, were certainly content to take for their model the common agricultural waggon of Germany. Indeed, Hungary seems to have played pioneer in this respect at a very early date. Von Ginzrot, in his work on early vehicles, gives an illustration of a closed passenger carriage which bears more than a superficial resemblance to the later coaches. "The body," says Thrupp, "is a disguised waggon; the tilt-top has two leather flaps to fall over the doorway, and the panels are of wicker-work." It would have been quite easy, he continues, to use such waggons, as had been the case long before, for passenger traffic, "by placing the planks across the sides, or suspending seats by straps from the sides"; and he further mentions an oil painting at Nuremberg, of two waggons "with carved and gilt standard posts both in front and behind the body" — an interesting stage in the transformation from rude cart to private coach. There is a detailed and technical description of these waggons in Thrupp's own book, but it will be enough here to notice that they were generally narrower at the bottom than at the top, as were the first coaches, and that the four wheels were nearly of the same size. Working from such a model, the Hungarian artificers produced a comparatively light, though large, four-wheeled carriage with some pretensions to grace of line, a roofed body, broad seats, and a side entrance. The body, however, was not completely enclosed

[20] It is probable that the closed carriage in which the Emperor Frederick III paid a visit to Frankfort in 1474 was one of these cotzi. Here the interesting point is that the Emperor's attendants, apparently for the first time, were relieved of the necessity of holding a canopy over His Majesty's head, except when he went to and returned from the Council Chamber.

by solid panels, which only took the place of the curtains at a later date. Carvings and other ornamentation followed on the owner's rank and taste. And towards the end of the sixteenth century, if not before, the actual body was suspended on straps or braces. There are preserved at Coburg and Verona one or two coach-bodies which show signs of the iron hoops by which they were hung. The earliest of these was built for Duke Frederick of Saxony in 1527, and Count Gozzadini, in a slim folio which he privately printed some sixty years ago, describes a coach-body built in 1549 which still shows traces of its heraldic ornamentation on the framework.

"This coach," says Thrupp, acting as the Count's translator, "was built under the direction of an Italian at Brussels, for the ceremony of the marriage of Alexander, the son of Octavius Farnese, Duke of Parma, with a Portuguese princess. The wedding took place in 1565 at Brussels. There were four carriages Flanders fashion [? charettes] and four coaches after the Italian fashion, swinging on leather braces. The chief, or state, coach is described as being in the most beautiful manner, with four statues at the ends, the spokes of the wheels like fluted columns. There were seraphims' heads at the end of the roof and over the doorway, and festoons of fruit in relief over the framing of the body. The coachman was supported by two carved figures of lions, two similar lions were at the hind wheel, and the leather braces that supported the body and the harness were embossed with heads of animals. The ends of the steps were serpents' heads. The whole of the wood and ironwork was covered with gold relieved with white. The coach was drawn by four horses, with red and white plumes of feathers, and the covering of the body and of the horses was gold brocade with knotted red silk fringe. The cushions of gold-embroidered stuff were perfumed with amber and musk, that infused the soul of all who entered the coach with life, joy, and supreme pleasure."

Truly a Southern notion!

What is apparently the oldest coach to be preserved practically intact is to be seen at Coburg. This coach was built for a particular occasion—the marriage of John, Elector of Saxony, in 1584. The body is long and ornate, and is hung from four carved standard posts surmounted by crowned lions. The wheels are large—four feet eight inches and five feet—and the roof is at a slightly higher level than the lions' heads. Mounting steps must have existed, but have been lost.

Elizabethan Carriages

From a Print by Hofnagel, 1582

Not unnaturally the advent of these coaches followed upon the commercial prosperity of each country. Germany seems to have imported a number of carriages from Hungary, and made others from Hungarian models, but even more prosperous than Germany at this time was Holland, which probably possessed more coaches than any other country in Europe. Here there would have been native designs to follow and improve upon, and, as I shall show in a moment, it was probably from the Netherlands that the first coach was imported into England. Antwerp, for instance, a superlatively rich city in the sixteenth century, is credited by Macpherson with having no less than five hundred coaches —and so five hundred scandals, according to the local philosophers—in 1560, at which date London had but two, and Paris no more than three. Of the French trio of *carosses*, as they were called, one was the Queen's property, a second belonged to the fashionable Diana of Poitiers, and the third had been built for the use of that corpulent noble who has already been mentioned. Some Italian towns possessed many, others none. There is preserved at the Musée Cluny in Paris a Veronese *carriole* built in the sixteenth century by Giovanna Batta Maretto, with panels painted by a distinguished artist of the time. Verona, indeed, seems to have had many coaches. But it was easily surpassed by Ferrara, which so early as 1509 is credited with the possession of no less than sixty coaches, the whole of these forming the Duke's procession on the occasion of a state visit from the Pope. And, as Thrupp points out, these sixty carriages were not litters or cars, as might be supposed, but coaches, for it is particularly mentioned by the historian that "the Duchess of Ferrara rode in a *litter*, and her ladies followed her in twenty-two *cars*." Spain had apparently no coaches until 1546, and here again there was considerable opposition to their use. Yet although England, France, and Spain seem to have been behind other countries in taking to the new carriages, all three possessed a flourishing, if not very large, coach-building trade before 1600.

Here, perhaps, we may consider the introduction of the coach into England in rather greater detail. "It is a doubtful question," remarks Taylor in his ill-natured way," whether the divell brought *Tobacco* into England in a *Coach*, or else brought a *Coach* in a fogge or mist of *Tobacco*." Apparently he had an equal dislike for both coach and tobacco. But although we owe to the water-poet such contemporary satirical writings on the subject as there are, he is not to be trusted as an historian. Taylor, indeed, is a very bad historian, not so much on account of his inability to see two sides of a question, as because, like many another poet, he has made of exaggeration a fine art, and allowed his memory to play second fiddle to his inclinations. It is to the worthy Stowe that we must turn for the facts. Stowe liked the coaches little better than did Taylor, but his training had made him exact, and we may take it for granted that he is more or less correct when he says that the first coach to be seen upon British roads belonged to the year 1555. Curiously enough, this is the date of the first General Highways Act. The preamble of this Bill stated that certain roads were "now both very noisesome and tedious to travel in and dangerous to all passengers and carriages [carts]." The local authorities were empowered to compel parishioners to give four days' work every year to the repairing of the roads, though how far such orders were carried out it would be impossible to say. The merit of actually introducing the coach is given by Stowe to Henry Manners, second Earl of Rutland, who caused one Walter Rippon to build him a carriage from some foreign, most probably Dutch, pattern. This Earl of Rutland had borne the Spurs at the coronation of Edward VI, and in 1547 had been made Constable of Nottingham Castle. He had received the French hostages in 1550 at the time of the treaty which followed on the loss of Boulogne. It is to be regretted that neither in his correspondence nor in the family account-books preserved at Belvoir is there mention of either Rippon or his coach. There is, indeed, the "Book of John Leek of riding charges carriages [carts] and forrene paymentes" in 1550, and another book compiled by Leek's successor, George Pilkington, in the following year, but all travelling entries concern only horses and the cartage of goods. In 1555 "George Lassells, Esquyer" was "Comptroller to the householde" and paid "to Edward Hopkynson for ij ryding roddes of bone for my Ladye and other thinges, xxij*d*," but there is no mention of any carriage for his Lordship's own use. What is more unfortunate is that there are no account-books of the Manners family between 1559 and 1585, and it is not until 1587, when a fourth Earl of Rutland was head of his house, that this significant entry occurs:—

"Coach, a newe, bought in London, xxxviij*li*.xiij*s*.ij*d*."

To go back to Rippon, it is not known who he was. He is supposed to have built a coach for Queen Mary in 1556, and in 1564 the first "hollow turning coach" with pillars and arches, for Queen Elizabeth, though precisely what is meant by a "hollow turning" coach it is difficult to conjecture. This same Rippon twenty-four years later built another coach for the Queen, which is described as "a chariot throne with foure pillars behind, to beare a crowne imperiale on the toppe, and before two lower pillars, whereon stood a lion and a dragon, the supporters of the armes of England." It cannot have been very comfortable, and Elizabeth seems to have preferred another coach brought out of Holland by one William Boonen, who about 1560 was made her coachman, a position he was still occupying at the end of the century. This Boonen was a Dutchman, whose wife is said to have introduced the art of starching into England, whence followed those huge ruffs so conspicuous in all the Elizabethan portraits. Boonen's coach could be opened and closed at pleasure. On the occasion of the Queen's passing through the town of Warwick, she had "every part and side of her coach to be opened, that all her subjects present might behold her, which most gladly they desired." This coach is described as "on four wheels with seven spokes, which are apparently bound round with a thick wooden rim secured by pegs. It is precisely such a vehicle," adds the anonymous historian in the *Carriage Builder's and Harness Maker's Art Journal*, "as is now (1860) used by the brewers, with a tilt over it, which opens in the centre on one side, and would contain half a dozen persons." On the other hand, one may safely assert that no brewer's cart was ever decorated in the same way, for the framing of Elizabeth's carriage was of wood carved in a shell pattern and gilded. "The whole composition," runs another account, "contains many beautiful curves. The shell-work creeps up to the roof, which it supports, and which is dome-shaped.... The roof is capped by five waving ostrich feathers, one at each corner, and the fifth on the centre of the roof, and springing from a kind of crown." The driver's seat was apparently a kind of movable stool, and two horses were used. Even this coach, however, of which there is a print by Hoefnagle, dated 1582, cannot have been very comfortable, and in 1568, when the French ambassador obtained an audience, Elizabeth was complaining of "aching pains" from being knocked about in a coach driven too fast a few days before. "No wonder," comments one historian, "that the great queen used her coach only when occasions of state demanded." Whenever possible, indeed, she used her horse. "When Queen Elizabeth came to Norwich,

1578," wrote Sir Thomas Browne a hundred years later, "she came on horseback from Ipswich, by the high road to Norwich, in the summer time; but she had a coach or two," he added, "in her trayne."

In the print just mentioned there is shown a second coach, which is perhaps a better example of the carriage of the period. One sees again its hearse-like appearance, though the top is broader than the bottom, and the body is partially enclosed; but there is one peculiarity which deserves particular mention. This was a small seat which projected on either side, between the wheels. It was known as the boot. Here sat the pages or grooms or the ladies in attendance. Taylor, of course, has his fling against it. The booted coach, he says, is like a perpetual cheater, wears "two Bootes and no Spurs, sometimes having two paire of Legs and one boote; and oftentimes (against nature) most preposterously it makes faire Ladies weare the boote; and if you note, they are carrried backe to backe like people surpriz'd by Pyrats to be tyed in that miserable manner, and throwne overboard into the Sea. Moreover, it makes people imitate Sea-crabs, in being drawne Side-wayes, as they are when they sit in the boote of the Coach." The boot, however, was already tending to disappear in Taylor's day. How it originated is not clear. It was always uncovered, whence followed much hardship, particularly if the weather was unfavourable. Nor can one think that it was very capacious. There is an early seventeenth-century pamphlet entitled *My Journie*, in which a stout old lady is put into the boot of a coach, and cannot move. When going uphill all the passengers are supposed to get out and walk, but the old lady, once settled, refuses to budge, and, indeed, cannot be extricated until the end of the journey. There is further mention of the discomfort in a boot in 1663, when Edward Barker, writing to his father, a Lancashire squire, complains of his troubles in the side seat. "I got to London," he says, "on Saturday last, my journey was noe ways pleasant, being forced to ride in the boote all the waye, ye company y^t came up wth mee were persons of greate quality as knightes and ladyes. My journeys expence was 30 s. This traval hath soe indisposed mee, y^t I am resolved never to ride againe in ye coatch. I am extreamly hot and feverish." The monstrous width of these early coaches followed, of course, on their projecting side seats, which only entirely disappeared when the coach had come to be completely enclosed and provided with glass windows.

It may be that the boot in process of time was metamorphosed into the large, deep, four-sided basket which was strapped to the back of public coaches in the seventeenth and eighteenth centuries, and, indeed, this basket seems to have been called the boot in eighteenth-century stage coaches. It was probably in such a basket-boot as this that Mr. Pepys put his great barrel of oysters, "as big as sixteen others," which was given him in 1664.

An interesting point in this connection is that those who travelled on the seatless and presumably most uncomfortable roof of a coach plying for hire, paid more for the privilege than did those who rode in the boot.

However greatly the chroniclers may differ as to the date of the actual introduction, and others besides Taylor disagree with Stowe, there seems no doubt that by 1585 many of the nobility and some wealthy commoners owned private coaches, and, indeed, certain enterprising tradesmen, as will appear, let other coaches on hire at so much per day.

> "After a while," says Stowe, "divers great ladies, with a great jealousy of the Queen's displeasure, made them coaches and rid them up and down the countries, to the great admiration of all the beholders, but then little by little they grew usual amongst the nobilitie and others of sort, and within twenty years became a great trade of coach-making."

Indeed, every one of any wealth was eager to possess them. A private coach settled any doubts as to your quality. It was a new fashion, a new excitement. "So a woman," says Quicksilver, the rake, in *Eastward Hoe*, "marry to ride in a coach, she cares not if she rides to her ruin. 'Tis the great end of many of their marriages." And again, in Ben Jonson's *Alchemist* it is said of the Countess that she

> "... has her pages, ushers
> Her six mares—
> Nay, eight!
> To hurry her through London, to the Exchange,
> Bethlem, the china-houses—
> Yes, and have
> The citizens gape at her, and praise her tires."

Even the plain country-folk seem to have been smitten with the new toy, for toy it was to them. "Has he ne'er a little odd cart," asks Waspe in *Bartholomew Fair*, "for you to make a coach on, in the country, with four pied hobby-horses?" Any shift for a coach, thought he, and no doubt voiced public opinion.

The first owners of coaches appear to have been those who had travelled abroad. So early as 1556, Sir Thomas Hoby, who had been our ambassador to France, possessed a coach and offered to lend it to the Lady Cecil. The account-book for 1573 of the Kytson family, of Hengrave, in Suffolk, mentions another early coach. "For my m^res [mistress's] coche, with all the furniture thereto belonging except horses—xxxiiij*li.*xiiij*s.* For the painting of my m^r and m^res armes upon the coche—ij*s.*vj*d.*" In 1579 the Earl of Arundel is said to have brought a coach into England from Germany, and this coach is interesting from the fact that certain historians have credited it with being the first coach in England. How such a tradition arose is not clear, but it may be that this German coach had certain features which more nearly approached those of the later Stuart, fully-enclosed, coaches. Further details are to be found in the Manners notebooks, and these afford a glimpse of the methods adopted by the coachmakers, not yet a large body, of the day. In the notebooks of Thomas Screven, 1596-97, after an item for twenty-eight shillings for three-quarters of "scarlet sleves and labelles for his L[ordship's] parlyament robes" comes another of six shillings "to my Lady Adeline's coachman," and one, just below, of greater interest:—

> "Item paid to Wm. Wright, coachmaker, in parte of xl*li.* for a coache now made, xx*li.*"

After that, in the 1598-99 book comes an item to "the Countess of South[ampton's] coachman that wayted on my Lord to Dertford, v*s.*" This suggests the growing popularity of the coach, more especially as there is another disbursement in the same year to the Countess of Essex's coachman. Then follow from November 25th, 1598, details of the expenses of the new coach for my Lord's own use—which apparently took considerable time to furnish.

> "Item for ij paire of new wheeles for the coache, tymber worke and iron work, and setting them on the axeltrees, iij*li*.xiij*s*.iiij*d*.; payntinge them in oyle colour, vj*s*.viij*d*.; a new pole for the horses to drawe by, ij*s*.vj*d*.; a paire of springe trees, iij*s*.iij*d*."

The provender bill for six horses is given, also an item "for setting up the coach horses at dyvers times at Walsingham Howse, iiij*s*.; at Hatton Howse, xij*d*.; at Baynardes Castle, ij*s*.; dressing and oyling the coach, ij*s*."; while the most necessary whip costs Mr. Screven twelve pence. Other payments are six shillings for two new bearing braces for the "double hanging" of the coach—here at any rate is definite mention of suspension, a fact which might suggest that, after all, either Rippon's or Lord Arundel's coach had been of the suspended type—four shillings for a long spring brace, two shillings and sixpence for a new "wynge," and sixteen pence for two "bearing raynes." The new coach, however, is not ready in time for his Lordship, who thereupon hires one with three horses to take him "to the Court at Nonesuch, 23, 24, and 25 of September, at xvj*s*. *per diem*." Meanwhile payments for his own coach continue. For four "skynnes of orange colour leather goate" he pays various sums; for the timber work, for more painting, for a covering in "black lether," and for making the "curtaynes, and setting on the firinge, and making the blew cloth cover" a sum of twenty-six pounds, nineteen shillings, is expended. Nor is this all. My Lord was evidently determined to make his coach as gorgeous as possible. Nine yards of "marygold coulour velvet for the seat and bed in the coach" were required, and each yard cost twenty-three shillings. The quilting for the bed cost forty shillings. In addition, there was a lace of "crymosin silk" and no less than "v elles of crymosin taffaty for curtaynes," costing three pounds fifteen shillings; also "9 yardes of blew clothe for a cover." Then, of great interest, comes the final entry:—

> "Item, paid to Ryly, embroderer, in full for embrodering iij sumpter clothes of crymosin with his L[ordship's] armes thereon at large, and vij otheres embrodered onely with great peacocks, with carsey for the garding and tasselles and frynge, 14 July, lxiij-j*li*."

Mr. Ryly was well paid for his work[21].

From such details it is possible to imagine what this and other coaches of the time were like. You figure a huge, gaudy, curtained apparatus with projecting sides and incomplete panels, large enough to contain a fair-sized bed, hung roughly from four posts, and capable of being dragged at little better than a snail's pace—"four-wheeled Tortoyses" Taylor calls them—along roads hardly worthy of the name. Twenty miles a day was considered good going. Says Portia, in the

[21] Taylor mentions in one place that "for the mending of the Harnesse, a Knights Coachman brought in a bill to his master of 25 pounds." He also says that the owners of coaches liked to match their horses if possible.

Merchant of Venice:—

> "... I'll tell thee all my whole device
> When I am in my coach, which stays for us
> At the park gate; and therefore haste away,
> For we must measure twenty miles to-day."

The coachman, as we learn from the water-poet, was "mounted (his fellow-horses and himselfe being all in a finery) with as many varieties of laces, facings, Clothes and Colours as are in the Rainebowe." Nor was he over-polite, particularly if the coach he drove was hired. In Jonson's *Staple of News* one of the pieces of mock-news to appear in the ideal paper concerns the fraternity:—

> "and coachmen
> To mount their boxes reverently, and drive
> Like lapwings with a shell upon their heads
> Through the streets."

They seem to have thought that their finery allowed them to treat the pedestrians with but scant respect. And no wonder these "way-stopping whirligigges," as Taylor calls the coaches, surprised the inhabitants. When one of them was seen for the first time, "some said it was a great Crab-shell brought out of *China*, and some imagin'd it to be one of the Pagan Temples in which the Cannibals adored the devill." For some time, indeed, the coaches must have given the common folk something to think about. A coach rumbling along brought them to their windows, just as the horseless carriage, centuries later, proved a similar attraction. There is a scene in *Eastward Hoe* which well illustrates this point.

Enter a Coachman in haste in 's frock, feeding.

Coach. Here's a stir when citizens ride out of town indeed, as if all the house were afire! 'Slight, they will not give a man leave to eat 's breakfast afore he rises.

Enter Hamlet, a footman, in haste.

Ham. What coachman? My lady's coach, for shame! her ladyship's ready to come down.

Enter Potkin, a tankard bearer.

Pot. 'Sfoot, Hamlet, are you mad? whither run you now?...

Enter Mrs. Fond and Mrs. Gazer.

Fond. Come, sweet mistress Gazer, let's watch here, and see my Lady Flash take coach.

Gazer. O' my word, here's a most fine place to stand in. Did you see the new ship launch'd last day, Mrs. Fond?

Fond. O God, and we citizens should lose such a sight!

Gazer. I warrant here will be double as many people to see her take coach, as there were to see it take water.

My lady's point of view is put forward by Lady Eitherside in *The Devil is an*

Ass. Says she:—

> "If we once see it under the seals, wench, then,
> Have with them for the great caroch, six horses,
> And the two coachmen, with my Ambler bare,
> And my three women; we will live, i' faith,
> The example of the town, and govern it.
> I'll lead the fashion still."

Contemporary references to coaches, however, are but scarce. The most important of these is Taylor's own *The World runnes on Wheeles: or, Oddes betwixt Carts and Coaches*, an amusing pamphlet written in prose and not in verse, because the author, as he says, was lame at the time of its composition, and because beyond the three words, broach, Roach, and encroach, he could find no suitable rhymes. Encroach, however, he thinks might have done, for that word, as he explains in his dedication to various companies likely to suffer from the importation of the coach, "best befits it, for I think never such an impudent, proud Intruder or Encroacher came into the world as a Coach is; for it hath driven many honest Families out of their Houses, many Knights to Beggers, Corporations to poverty, Almesdeedes to all misdeedes, Hospitality to extortion, Plenty to famine, Humility to pride, Compassion to oppression, and all Earthly goodnes to an utter confusion." To the cart he does not object, but for the "hyred Hackney-hell-carts" he cannot find sufficient abuse. His arguments in favour of carts as against coaches are certainly novel, if not entirely convincing as coming from a waterman well used to live passengers himself.

> "And as necessities and things," he says, "whose commodious uses cannot be wanted, are to be respected before Toyes and trifles (whose beginning is Folly, continuance Pride, and whose End is Ruine) I say as necessity is to be preferred before superfluity, so is the *Cart* before the *Coach*; For Stones, Timber,.Corne, Wine, Beere, or any thing that wants life, there is a necessity they should be carried, because they are dead things and cannot go on foot, which necessity the honest *Cart* doth supply: But the *Coach*, like a superfluous bable, or uncharitable Miser, doth seldom or never carry or help any dead or helplesse thing; but on the contrary, it helps those that can helpe themselves ... and carries men and women, who are able to goe or run; *Ergo*, the *Cart* is necessary, and the *Coach* superfluous."

In fact, the coach, according to poor Taylor, is directly responsible for every calamity from which the country has suffered since its introduction. Leather has become dearer, the horses in their traces are being prostituted, and there is a "universal decay of the best ash-trees."

> "A Wheele-wright," he continues, "or a maker of Carts, is an ancient, a profitable and a Trade, which by no meanes can be wanted: yet so poore it is, that scarce the best amongst them can hardly ever attaine to better than a Calves skin fate, or a piece of

beefe and Carret rootes to dinner on a Sunday; nor scarcely any of them is ever mounted to any Office above the degree of a Scavenger, or a Tything-man at the most. On the contrary, your Coachmakers trade is the most gaine-fullest about the Towne, they are apparelled in Sattens and Velvets, and Masters of their Parish, Vestry-men, who fare like the Emperors *Heliogabalus* or *Sardanapalus*, seldome without their Mackroones, Parmisants, Jellies and Kickshawes, with baked Swannes, Pasties hot, or cold red Deere Pyes, which they have fro their Debtor Worships in the Country: neither are these Coaches onely thus cumbersome by their Rumbling and Rutting, as they are by their standing still, and damming up the streetes and lanes, as the Blacke Friers, and divers other places can witnes, and against Coachmakers doores the streets are so pestered and clogg'd with them, that neither man, horse or cart can passe for them; in so much as my Lord Maior is highly to bee commended for his care in their restraint, sending in February last, many of them to the Courtes for their carelessnesse herein."

In another work of Taylor's, *The Thiefe*, there is a passage of equal interest:—

"Carroaches, Coaches, Jades and Flanders Mares
Do rob us of our shares, our wares, our Fares:
Against the ground we stand and knock our heeles,
Whilest all our profit runs away on wheeles;
And whosoever but observes and notes,
The great increase of Coaches and of Boats,
Shall finde their number more than e'r they were
By halfe and more within these thirty yeeres.
Then watermen at Sea had service still,
And those that staid at home had worke at will:
Then upstart Helcart-Coaches were to seeke,
A man could scarce see twenty in a weeke,
But now I thinke a man may daily see,
More than the Whirries on the *Thames* can be.
When Queen *Elizabeth* came to the Crowne,
A Coach in *England*, then was scarcely knowne,
Then 'twas as rare to see one, as to spy
A Tradesman that had never told a lye."

It will be seen from the first of these lines, that a difference is made between the coach and the caroche (carroch or carroache). On this point there is a definite statement in the Elizabethan play *Tu Quoque*:—

"Prepare yourself to like this gentleman,
Who can maintain thee in thy choice of gowns,
Of tires, of servants, and of costly jewels;
Nay, for a need, out of his easy nature,
May'st draw him to the keeping of a coach
For country, and carroch for London."

This, too, is borne out by the speech of Lady Eitherside already quoted. Many servants were needed for the carroch. Massinger speaks of one being drawn by six Flanders mares, and having its coachman, groom, postilion, and footman, to look after it. "These carroaches," says Croal[22] "were larger and clumsier" than the coaches, "but were considered more stately." Taylor speaks of the town Vehicle as "a mere Engine of Pride," and gives a rather ludicrous account of some common women who had hired one of them to go to "the Greene-Goose faire at *Stratford* the *Bowe*." The occupants of this carroch "were so be-madam'd, be-mistrist, and Ladified by the beggers, that the foolish Women began to swell with a proud Supposition or Imaginary greatnes, and gave all their mony to the mendicanting Canters."

Poor Taylor! He felt very deeply on the question of these new coaches which were to put an end once and for all time to his trade. He must have felt that Henry of Navarre's assassination in 1610 would never have taken place but for that monarch's affection for his coach; yet in spite of his deep hatred, he was once prevailed upon to ride inside one of them. "It was but my chance" he records, "once to bee brought from Whitehall to the Tower in my Master Sir William Waades Coach, and before I had been drawn twenty yardes, such a Timpany of Pride puft me up, that I was ready to burst with the winde chollicke of vaine-glory. In what state I would leane over the boote, and looke, and pry if I saw any of my acquaintance, and then I would stand up vailing my Bonnet."

It almost looks as though he had enjoyed his ride!

---

[22] *A Book about Travelling, Past and Present.* Thomas Croal. London, 1877.

CHAPTER THE FOURTH
INTERLUDE OF THE CHAIR

"I love sedans, cause they do plod
And amble everywhere,
Which prancers are with leather shod,
And ne'er disturb the eare.
Heigh doune, derry derry doune,
With the hackney Coaches doune,
Their jumping make
The pavements shake,
Their noise doth mad the toune."

Ancient Ballad.

J UST as the horse-litter gave way before the coach, so the coach, not long after its appearance, found a serious rival in the man-drawn litter or Sedan chair. When or where this chair came from, or who brought it into use once again, is not known. That Sedan itself was the first place to adopt this chair may be true—the analogy already mentioned holding good—but beyond a few half-serious words in a curious seventeenth-century pamphlet to be quoted in a little, there is no positive evidence whatever. Several writers, indeed, assert that Sedan had nothing to do with the chair for ever associated with its name, but in that tantalising manner which is unfortunately characteristic of former times, omit to state their reason. It has been suggested that *sedan* was the name of the cloth with which the chair was lined, but if this were so, the cloth most probably took its name from the chair it adorned. But wherever it was first made it is reasonable to suppose that the narrowness of the streets made a smaller vehicle than either coach or horse-litter convenient.

The earliest chair, other than those ancient *lecticæ* and φορεῖα mentioned in the first chapter, appears to have belonged to the Emperor Charles V, in the first half of the sixteenth century. This, indeed, does bear some resemblance to the common conception of a *chair*, but the first Sedans of some fifty years later resembled nothing so much as a modern dog-kennel provided with two poles. A more unsociable apparatus was surely never built, and yet its almost immediate popularity is easily explained. With the urban streets not yet properly paved and the eternal jolting of the coach, to the accompaniment of such a clatter as must have made speech almost impossible, anything in the nature of a conveyance that made at once for physical comfort and comparative silence would have been favourably received.

There is mention of a chair being shown in England in 1581—just at the time when the country was beginning to show an interest in carriages—but it was not until after the death of Elizabeth that such a novelty was seen in the streets of London. You are not wholly surprised, moreover, to hear that the innovation was due to Buckingham, that apostle of luxury, who probably first saw the chair on his visit to Spain with Prince Charles. Indeed the Prince is supposed to have brought back three of them with him.

At first, of course, there was opposition.

> "Every new thing the People disaffect," wrote Arthur Wilson, the historian, "They stumble sometimes, at the *action* for the *person*, which rises like a little *cloud* but soon after vanishes. So after, when *Buckingham* came to be carried upon Men's shoulders the *clamour* and the *noise* of it was so extravagant that the People would rail on him in the Streets, loathing that Men should be brought to as servile a condition as Horses. So irksome is every little new impression that breaks an old *Custom* and rubs and grates against the *public humour*. But when Time had made these Chairs common, every loose *Minion* used them, so that that which got at first so much scandal was the means to convey those privately to such places where they might give much more. Just like *long hair*, at one time described as abominable, at another time approved as beautiful. So various are the *fancies* of the *times*!"

It is to be noticed that Buckingham, according to this account, was carried upon men's shoulders. This was the case at first, but such a mode was speedily changed for that of hand-poles—at once safer and more comfortable for the occupant, and certainly more convenient for the men.[23]

John Evelyn disagrees with Wilson and ascribes the introduction of the chair into England to Sir Saunders Duncombe, a Gentleman-Pensioner knighted by James I in Scotland in 1617, who enjoyed Buckingham's patronage. In his Diary for 1645, he writes of the Neapolitans: "They greatly affect the Spanish gravity in their habit; delight in good horses; the streets are full of gallants on horseback, in coaches and sedans, from hence brought first into England by Sir Saunders Duncombe." Undoubtedly Duncombe was responsible for the great popularity of the chair in England, and for a time held a monopoly in such chairs as could be had for hire, but it may be that Buckingham suggested this monopoly in the first place, after the temporary opposition to their use had been overcome. Which rather suggests that Spain was actually the first country where they were used, though this is mere conjecture.

In the meantime much was happening to the coaches. They were increasing

---

[23] So Massinger in *The Bondman* says:—

> "For their pomp and ease being borne
> In triumph on men's shoulders."

enormously in number, not only those privately owned, but also those hired out by the day. These latter soon became known as hackney-coaches.[24] They seem to have been put on the streets as early as 1605, but "remained in the owner's yards until sent for." In 1633 the Strand was chosen as the first regular stand for such coaches by a Captain Bailey, one of the pioneers of the movement.

> "I cannot omit to mention," writes Lord Stafford, "any new thing that comes up amongst us though ever so trivial. Here is one Captain Bailey, he hath been a sea captain, but now lives on land about this city where he tries experiments. He hath erected, according to his ability, some four hackney coaches, put his men in livery and appointed them to stand at the Maypole in the Strand, giving them instructions at what rate to carry men into several parts of the town where all day they may be had. Other hackney men veering this way, they flocked to the same place and performed their journeys at the same rate, so that sometimes there is twenty of them together, which dispose up and down, that they and others are to be had everywhere, as watermen are to be had at the waterside. Everybody is much pleased with it, for whereas before coaches could be had but at great rate"—one recalls the prices paid by Lord Rutland a few years before—"now a man may have one much cheaper."

Most of these coaches that were put on to the streets seem to have been old and disused carriages belonging to the quality. Many of them still bore noble arms, and, indeed, it would seem that when the hackneys were no longer disused noblemen's carriages, the proprietors found it advisable to pretend that they were. Nearly every hansom and four-wheeled cab at the end of the nineteenth century bore some sort of coronet on its panels.

The drivers of these first hackneys wore large coats with several capes, one over the other, for warmth. London, however, seems to have been the only town in which they were to be seen. "Coaches," wrote Fynes Morison in 1617, "are not to be hired anywhere but in London. For a day's journey a coach with two horses is let for about 10s. a day, or 15s. with three horses, the coachman finding the horses' feed." From the same author it would appear that most travellers still doggedly kept to their horses, and indeed, in some counties a horse could be hired for threepence a day, an incredibly small sum. "Carriers," he also records, "have long covered waggons in which they carry passengers too and fro; but this kind of journeying is very tedious; so that none but women and people of inferior condition travel in this sort." These were the stage-waggons which in due course gave rise to the stage-coaches, which in their turn were superseded by the mail-coaches.

A similar movement in France gave rise to the *fiacres*, so called from the sign of St. Fiacre, which adorned one of the principal inns in Paris, in front of which the public coaches stood. In Scotland, too, one Henry Andersen, a native of Pomer-

---

[24] The word hackney, possibly derived from the old French *Haquenée*, was the natural word to be used for a public coach, it being merely a synonym, used by Shakespeare and others, for *common*.

ania, had in 1610 been granted a royal patent to provide public coaches in Scotland, and for some years ran a service between Edinburgh and Leith. England had yet to follow Andersen's example, but the hackneys were increasing so rapidly in London that in 1635 a proclamation was issued to suppress them. And it is to be noticed that Taylor's diatribes were directed more particularly against these public conveyances than against the privately owned carriages, which, after all, could hardly affect his trade. The proclamation was as follows:—

> "That the great numbers of Hackney Coaches of late time seen and kept in London, Westminster, and their Suburbs, and the general and promiscuous use of Coaches there, were not only *a great disturbance to his Majesty, his dearest Consort the Queen, the Nobility, and others of place and degree, in their passage through the Streets;* but the Streets themselves were so pestered, and the pavements so broken up, that the common passage is thereby hindered and more dangerous; and the prices of hay and provender and other provisions of stable, thereby made exceeding dear: Wherefore We expressly command and forbid, That, from the feast of St. John the Baptist next coming, no Hackney or Hired Coach, be used or suffered in London, Westminster, or the Suburbs or Liberties thereof, excepting they be to travel at least three miles out of London or Westminster, or the Suburbs thereof. And also, that no person shall go in a Coach in the said Streets, except the owner of the Coach shall constantly keep up Four able Horses for our Service, when required."

It is dated January 19th, 1635/6, and must have had a considerable, if temporary, effect, for as Samuel Pegge points out in his unfinished manuscript on the early use of coaches[25] it could not "operate much in the King's favour, as it would hardly be worth a Coach-master's while to be at so great a contingent charge as the keeping of four horses to be furnished at a moment's warning for His Majesty's occasional employment."

It was then that Sir Saunders Duncombe obtained his monopoly, and, of course, everything was in his favour. The actual patent granted to him belongs to the previous year, but the two are approximately contemporary. From a letter written in 1634 to Lord Stafford, it appears that Duncombe had in that year forty or fifty chairs "making ready for use." Possibly the whole thing was worked up by Buckingham and his satellites. Duncombe's patent gave the enterprising knight the right "to put forth and lett for hire" the new chairs for a term of fourteen years. In his petition he had explained that "in many parts beyond the seas, the people there are much carried in the Streets in Chairs that are covered; by which means very few Coaches are used amongst them." And so Duncombe was allowed to "reap some fruit and benefit of his industry," and might "recompense himself of the costs, charges, and expences" to which he had, or said he had, been put.

For two years these covered chairs held the advantage, and indeed seem to have been exceedingly popular. There is a most amusing pamphlet, which I have

[25] *Curialia Miscellanea.* Samuel Pegge, F.S.A. London, 1818.

already mentioned, "printed by Robert Raworth, for John Crooch," in 1636, entitled *Coach and Sedan pleasantly disputing for Place and Precedence, the Brewer's Cart being Moderator.* It is signed "Mis-amaxius," and is dedicated "to the Valorous, and worthy all title of Honor, Sr Elias Hicks." "Light stuffe," the author calls it, and tells us that he is "no ordinary Pamphleteer ... onely in Mirth I tried what I could doe upon a running subject, at the request of a friend in the *Strand*: whose leggs, not so sound as his Judgement, enforce him to keepe his Chamber, where hee can neither sleepe or studie for the clattering of *Coaches*." It is an interesting little production, both for its own whimsicalities and for the sidelights it affords into the town's views on the subject of vehicles at the time. It starts with the cuckoo warning the milkmaids of Islington to get back to *Finsburie*. The writer, accompanied by a Frenchman and a tailor, walks back to the city, and in a narrow street comes across a coach and a sedan quarrelling about which of them is to "take the wall."

> "Wee perceived two lustie fellowes to justle for the wall, and almost readie to fall together by the eares, the one (the lesser of the two) was in a suite of greene after a strange manner, windowed before and behind with *Isen-glasse*, having two handsome fellowes in greene coats attending him, the one went before, the other came behind; their coats were lac'd down the back with a greene-lace sutable, so were their halfe sleeves, which perswaded me at first they were some cast suites of their Masters; their backs were harnessed with leather cingles, cut out of a hide, as broad as *Dutch*-collops of *Bacon*.

> "The other was a thick burly square sett fellow, in a doublet of Black-leather, Brasse-button'd down the brest, Backe, Sleeves, and winges, with monstrous wide bootes, fringed at the top, with a net fringe, and a round breech (after the old fashion) guilded, and on his back-side an Atcheivement of sundry Coats in their propper colors, quarterd with Crest, Helme and Mantle, besides here and there, on the sides of a single Escutchion or crest, with some Emblematicall *Word* or other; I supposed, they were made of some Pendants, or Banners, that had beene stollen, from over some Monument, where they had long hung in a Church.

> "Hee had onely one man before him, wrapt in a red cloake, with wide sleeves, turned up at the hands, and cudgell'd thick on the backe and shoulders with broad shining lace (not much unlike that which Mummers make of strawe hatts) and of each side of him, went a Lacquay, the one a French boy, the other Irish, all sutable alike: The *Frenchman* (as I learned afterward) when his Master was in the Countrey, taught his lady and his daughter *French*: Ushers them abroad to publicke meetings, and assemblies, all saving the Church whither shee never came: The other went on errands, help'd the maide to beate Bucks, fetch in water, carried up meate, and waited at the Table."

The writer attempts mediation, and his offer is favourably received. The combatants explain who they are. The burly fellow speaks first:—

> "My name Sir (quoth hee) is *Coach*, who am a Gentleman of
> an anciente house, as you may perceive by my so many quarter'd
> coates, of *Dukes*, *Marquises*, *Earles*, *Viscounts*, *Barons*, Knights, and
> Gentlemen, there is never a Lord or Lady in the land but is of my
> acquaintance; my imployment is so great, that I am never at qui-
> et, day or night; I am a Benefactor to all Meetings, Play-houses,
> Mercers shops, Taverns, and some other houses of recreation....
> This other that offers me the wrong, they call him Mounsier *Sedan*,
> some Mr. *Chair*, a Greene-goose hatch'd but the other day ... and
> whereas hee is able with all the helpe and furtherance hee can
> make and devise, to goe not above a mile in an houre; as grosse as
> I am, I can runne three or foure in halfe an houre; yea, after din-
> ner, when my belly is as full as it can hold (and I may say to you)
> of dainty bitts too."

Whereupon the sedan chimes in:—

> "Sir, the occasion of our difference was this: Whether an emp-
> tie Coach, that has a Lords head painted Coate and Crest, as Lion,
> Bull, Elephant, &c. upon it without, might take the wall of a *Se-*
> *dan* that had a knighte alive within it." I confess, he goes on to
> say, I am "a meere stranger, till of late in *England*; therefore, if
> the Law of Hospitalitie be observed (as *England* hath beene ac-
> counted the most hospitable kingdome of the World,) I ought to
> be the better entertained, and used, (as I am sure I shall) and find
> as good friends, as Coach hath any, it is not his bigge lookes, nor
> his nimble tongue, that so runnes upon wheeles, shall scare mee;
> hee shall know that I am above him both in esteeme, and digni-
> tie, and hereafter will know my place better.... Neither, I hope,
> will any thinke the worse of mee, for that I am a Forreiner; hath
> not your Countrey Coach of England been extreemly enriched by
> strangers?"

Indeed, all your luxuries, he continues, are foreign, your perfumes are Italian, and
your perukes made in France.

For some time it seems that Sedan is getting the best of it. Whereas the coach,
he argues, has to wait out in the cold streets often for hours at a time, he is many
times admitted into the privacy of my Lady's chamber, where he is rubbed clean
both within and without. "And the plain troath is," he concludes, "I will no longer
bee made a foole by you ... the kenell is your naturall walke." At this moment a
carman appears and supports the sedan. Coaches, he says, keep the town awake,
endanger the lives of children, and, particularly in the suburbs, "be-dash gentlemen's gowns." There then follows a curious piece of dialogue between Sedan and
Powel, a Welshman, one of his attendants:—

> "*Sedan.* We have our name from *Sedanum*, or *Sedan*, that fa-
> mous Citie and Universitie, belonging to the Dukes of *Bevillon*,

and where hee keepes his Court."

> "*Powel.* Nay, doe you heare mee Master, it is from *Sedanny,* which in our British language, is a brave, faire, daintie well-favoured Ladie, or prettie sweete wench, and wee carrie such some time Master...."

Most of the morning is wasted by such desultory talk, and the street becomes blocked. There comes on the scene a waterman, who, of course, is equally antagonistic to both, and would throw coach and sedan into the Thames if he were not afraid of blocking the stream, and so bringing harm to himself. There follows him a country farmer, who thinks the sedan the honester and humbler of the two, but really knows very little about it. "I heare no great ill of you," he is good enough to say, but is bound to add, "I have had no acquaintance with your cowcumber-cullor'd men." Yet in the country he has in his way tried a sedan-chair, which is a "plaine wheele-barrow," just as his cart is his coach "wherein now and then for my pleasure I ride, my maides going along with me." But if they both come to Lincolnshire, the sedan, he thinks, will receive a warmer welcome than the coach.

After him comes a country vicar who has no hesitation in accusing the coach of all sorts of robberies. Soon, he cries, you will be "turned off." You never cared for church, and indeed, during service, you disturb everybody rumbling your loudest outside. Also you are so set up that you will never give place "either to cart or carre." A surveyor is less personal than the vicar, but has little good to say of the coach, although he agrees with most of the others that for a nobleman of high rank, it is something of a necessity.

Finally the brewer appears and speedily puts an end to the wrangle.

> "With that, comes up unto us a lustie tall fellow, sitting betweene two mōstrous great wheeles, drawne by a great old jade blinde of an eie, in a leather pilch, two emptie beere-barrels upon a brewer's slings besides him, and old blew-cap all bedaub'd, and stincking with yest.... My name is *Beere-cart*, quoth hee, I came into England in *Henry* the Seventh's time."

And the decision of the cart is, of course, that both coach and sedan shall give way to *him*. They are both to exercise great care, and the sedan is to have the wall. And he adds, turning to the smaller vehicle, a sentence which it is difficult to understand.

> "You shall never," he says, "carrie Coachman againe, for the first you ever carried was a Coachman, for which you had like to have sufferd, had not your Master beene more mercifull."

Such quarrels were very frequent, not only at this time, but right on through the eighteenth century. Swift in one of his letters to Stella mentions an accident due to the carelessness of a chairman. "The chairman that carried me," he says, "squeezed a great fellow against a wall, who wisely turned his back, and broke one of the side glasses in a thousand pieces. I fell a scolding, pretended I was like to be cut to pieces, and made them set down the chair in the Park, while they picked out the bits of glasses: and when I paid them, I quarrelled still, so they dared not grumble, and I came off for my fare: but I was plaguily afraid they would have said, God bless your honour, won't you give us something for our glass?"

Swift was the author of an amusing satire on the same subject, wherein coach and sedan were no better friends than of old.

A CONFERENCE BETWEEN SIR HARRY PIERCE'S CHARIOT AND MRS. D. STOPFORD'S CHAIR

CHARIOT

"My pretty dear Cuz, tho' I've roved the town o'er,
To dispatch in an hour some visits a score;
Though, since first on the wheels, I've been everyday
At the 'Change, at a raffling, at church, or a play;
And the fops of the town are pleased with the notion
Of calling your slave the perpetual motion;—
Though oft at your door I have whined [out] my love
As my knight does grin his at your Lady above;
Yet, ne'er before this though I used all my care,
I e'er was so happy to meet my dear Chair;
And since we're so near, like birds of a feather,
Let's e'en, as they say, set our horses together.

CHAIR

"By your awkward address, you're that thing which should carry,
With one footman behind, our lover Sir Harry.
By your language, I judge, you think me a wench;
He that makes love to me, must make it in French.
Thou that's drawn by two beasts, and carry'st a brute,
Canst thou vainly e'er hope, I'll answer thy suit?
Though sometimes you pretend to appear with your six,
No regard to their colour, their sexes you mix:
Then on the grand-paw you'd look very great,
With your new-fashion'd glasses, and nasty old seat.
Thus a beau I have seen strut with a cock'd hat,
And newly rigg'd out, with a dirty cravat.
You may think that you make a figure most shining,
But it's plain that you have an old cloak for a lining.
Are those double-gilt nails? Where's the lustre of Kerry,
To set off the Knight, and to finish the Jerry?
If you hope I'll be kind, you must tell me what's due
In George's-lane for you, ere I'll buckle to.

CHARIOT

"Why, how now, Doll Diamond, you're very alert;
Is it your French breeding has made you so pert?
Because I was civil, here's a stir with a pox:

> Who is it that values your — — or your fox?
> Sure 'tis to her honour, he ever should bed
> His bloody red hand to her bloody red head.
> You're proud of your gilding; but I tell you each nail
> Is only just tinged with a rub at her tail;
> And although it may pass for gold on a ninny,
> Sure we know a Bath shilling soon from a guinea.
> Nay, her foretop's a cheat; each morn she does black it,
> Yet, ere it be night, it's the same with her placket.
> I'll ne'er be run down any more with your cant;
> Your velvet was wore before in a mant,
> On the back of her mother; but now 'tis much duller, —
> The fire she carries hath changed its colour.
> Those creatures that draw me you never would mind,
> If you'd but look on your own Pharaoh's lean kine;
> They're taken for spectres, they're so meagre and spare,
> Drawn damnably low by your sorrel mare.
> We know how your lady was on you befriended;
> You're not to be paid for 'till the lawsuit is ended:
> But her bond it is good, he need not to doubt;
> She is two or three years above being out.
> Could my Knight be advised, he should ne'er spend his vigour
> On one he can't hope of e'er making *bigger*."

Gay seems to have shared the watermen's disgust at both coach and sedan.

> "Boxed within the chair, contemn the street
> And trust their safety to another's feet,"

he says of those willing to use the chair. In another place he is comparing the
two: —

> "The gilded chariots while they loll at ease
> And lazily insure a life's disease;
> While softer chairs the tawdry load convey
> To court, to *White's*, assemblies or the play."

Elsewhere he exhorts the pedestrian to assert his rights: —

> "Let not the chairman, with assuming stride,
> Press near the wall, and rudely thrust thy side;
> The laws have set him bounds; his servile feet
> Should ne'er encroach where posts defend the street."

By this time, however, many changes in the chairs had taken place. They seem
to have been introduced into Paris in 1617 by M. de Montbrun, though unfortunately from whence this gentleman brought them we are nowhere informed. They
were called *chaises à porteurs*. Possibly English and French chairs were at first quite
similar to each other in appearance — square boxes with a pent-house — but in the
middle of the century — in Paris, at any rate, they became far more elegant in form,
and began to be ornamented and richly upholstered. Some of them resembled, in

shape, the body of the modern hansom-cab. This was particularly the case with a new carriage, introduced about 1668, called the *brouette* (wheelbarrow), *roulette*, or *vanaigrette*, which was merely a sedan upon two wheels. It was drawn in the usual way by a man, and was an early form of that vehicle which still survives in the East as the jin-rick-shaw. The brouette held but one person, its wheels were large, and its two poles projected some way in front. One Dupin was apparently the only person to manufacture them, and after his first experiments he applied "two elbow-springs beneath the front, and attached them to the axle-tree by long shackles, the axle-tree working up and down in a groove beneath the inside-seat." This improvement is of more than ordinary interest in so far as it is the first mention of steel springs to carriages. In the ordinary coaches these steel springs were first applied beneath the bottom of the body. They were probably formed out of a single piece of metal.

In the case of the brouette there was the usual opposition—this time from the proprietors of the ordinary sedans—but although a temporary prohibition was made, the brouette triumphed, and in 1671 was a common sight in the streets of Paris. It was not very suitable for decoration. As one French writer remarks, it was enough if the machine were solidly constructed. The brouette had windows at the sides and a small support in front of the wheels to allow the carriage to maintain its proper position when not held up by an attendant.

The brouette does not seem to have come immediately to England, though in the eighteenth century there was a *sedan cart*, similar in appearance to it, to be seen in London. On the other hand, the ordinary sedans were rapidly gaining in popularity, and maintained that popularity right through the reigns of the first three Georges.

In appearance they became rather more graceful towards the middle of the century, though less so in later days. The public chairs were generally made of black or dark green leather, ornamented with gold "beading," the frame and roof, which had a double slope, being of wood, as was also the small square window-frame. Private chairs, however, could be as gorgeous as the owner pleased, though in this respect continental chairs far surpassed our own. At Paris are shown two magnificent chairs which belonged to Louis XV.

> "These," says Croal, "have glass windows in side and front, through which the sumptuous lining of crimson velvet is discernible. The outside is beautifully painted and gilt, and though now somewhat faded, the splendour of the vehicles can be imagined, even in their decay. The gorgeously attired king within, or it might be the queen or some reigning favourite, would be attended by a gay escort of gentlemen of the court, with a crowd of bearers and lacqueys, not to speak of armed guards, whose liveries probably equalled in grandeur the courtly habits of the greater men who surrounded the royal chair."

Neapolitan Sedan Chair Early Sixteenth Century

(At South Kensington)

At South Kensington a private English chair of about 1760 is shown, "rather handsomely ornamented in ormolu, the sides being divided into four panels, but without windows. In form," continues Croal, "the chair may be described as 'carriage-bodied,' not being, as the later chairs, square at the bottom. At the two front corners heavy tassels are hung, and through the door in front it can be seen that the interior lining is of figured damask. The bearing rings through which the poles passed are of brass." This, however, cannot compare with an Italian noble-man's large conveyance of the early eighteenth century which shows a profusion of gold filigree work on the roof that calls to mind nothing so strongly as a Buszard wedding-cake. It belonged to a member of the Grand Ducal family of Tuscany, by whom it was used on baptismal occasions. Here, besides the gilt work on the roof, there is a medallion-painting of figures in antique costume over the door. The walls are painted a pale French grey "with elaborately carved mouldings round the panels, with groups of flowers painted in the middle. The interior is lined with satin corresponding to the painting outside, being in gold and colours upon a pale ground."

The chairmen do not seem to have been a particularly agreeable lot of fellows. In London they were generally Irish or Welsh. They were often drunk, often care-less, and nearly always uncivil. Says Gay:—

> "The drunken chairman in the kennel spurns,
> The glasses shatter, and his charge o'erturns."

In Edinburgh, however, where there were ninety chairs in 1738, the chairmen were Highlanders and rather more civil. "An inhabitant of Edinburgh," says Hugh Arnot in his history of that city (1789), "who visits the metropolis can hardly sup-press his laughter at seeing the awkward hobble of a street chair in the city of Lon-don." We learn from Markland that in 1740 a chair in Edinburgh could be hired for four shillings a day or twenty shillings a week.[26] In London, according to George Selwyn, you could be carried three miles for a shilling.[27] In Edinburgh, again, where chairs were used at a later date than anywhere in England, rules were made for the public convenience in 1740, the most interesting of these being one which forbade a soldier in the service of the city guard to carry a chair at any time. By 1789 their numbers had increased to 238, including fifty privately owned.

Scattered mention of them occurs amongst British authors. Steele, in one of his *Tatler* papers, proposes to levy a tax upon them, and regrets that the sumptuary laws of the old Romans have never been revived. The chairmen, or "slaves of the rich," he says, "take up the whole street, while we Peripatetics are very glad to

---

[26] Which was about the same sum that Defoe had to pay in London earlier in the cen-tury. "We are carried to these places [the coffee-houses]," he wrote in 1702, "in chairs which are here very cheap—a guinea a week, or a shilling per hour; and your chairmen serve you for porters, to run on errands, as your gondoliers do at Venice."

[27] cf.

> "With chest begirt by leathern bands,
> The chairman at his corner stands;
> The poles stuck up against the wall
> Are ready at a moment's call.
> For customers they're always willing
> And ready aye to earn a shilling."

Echoes of the Street.

watch an opportunity to whisk across a passage, very thankful that we are not run over for interrupting the machine, that carries in it a person neither more handsome, wise, nor valiant, than the meanest of us."

Matthew Bramble in *Humphrey Clinker* is made to draw a wretched picture of the chairs which abounded in Bath at the middle of the century:—

> "The valetudinarian," he writes, "is carried in a chair, betwixt the heels of a double row of horses, wincing under the curry-combs of grooms and postilions, over and above the hazard of being obstructed or overturned by the carriages which are continually making their exit or their entrance. I suppose, after some chairmen shall have been maimed, and a few lives lost by those accidents, the corporation will think in earnest about providing a more safe and commodious passage.... If, instead of the areas and iron rails, which seem to be of very little use, there had been a corridor with arcades all round, as in Covent Garden, the appearance of the whole would have been more magnificent and striking; those arcades would have afforded an agreeable covered walk, and sheltered the poor chairmen and their carriages from the rain, which is here almost perpetual. At present the chairs stand soaking in the open street from morning to night, till they become so many boxes of wet leather, for the benefit of the gouty and rheumatic, who are transported in them from place to place. Indeed, this is a shocking inconvenience, that extends over the whole city; and I am persuaded it produces infinite mischief to the delicate and infirm. Even the close chairs, contrived for the sick, by standing in the open air, have their fringe linings impregnated, like so many sponges, with the moisture of the atmosphere."

It was to Bath that Princess Amelia was carried in a sedan by eight chairmen from St. James's, in April, 1728. This must easily have been the longest, and, so far as the chairmen were concerned, the most wearisome journey ever performed by a chair.

John Wilkes mentions in one of his letters to his daughter that he ascended Mont Cenis in a chair "carried by two men and assisted by four more." "This," he says, "was not a sedan chair, but a small wicker chair with two long poles; there is no covering of any kind to it." Such open chairs seem to have been very uncommon, and were, I imagine, unknown in England. Some, however, had more glass than others, and their size fluctuated. Fashionable ladies must have found a difficulty in getting into a public chair of the ordinary size at the time of the large hoop petticoat, and there is a satiric print, dated 1733, which shows a lady thus attired, being hauled out through the opened roof of one with ropes and pulleys. Similarly, when forty or fifty years later the head-dress of the women became so enormous, a ludicrous print appeared showing a patent arrangement whereby the roof of a chair could be raised on rods to as great a height as was required.

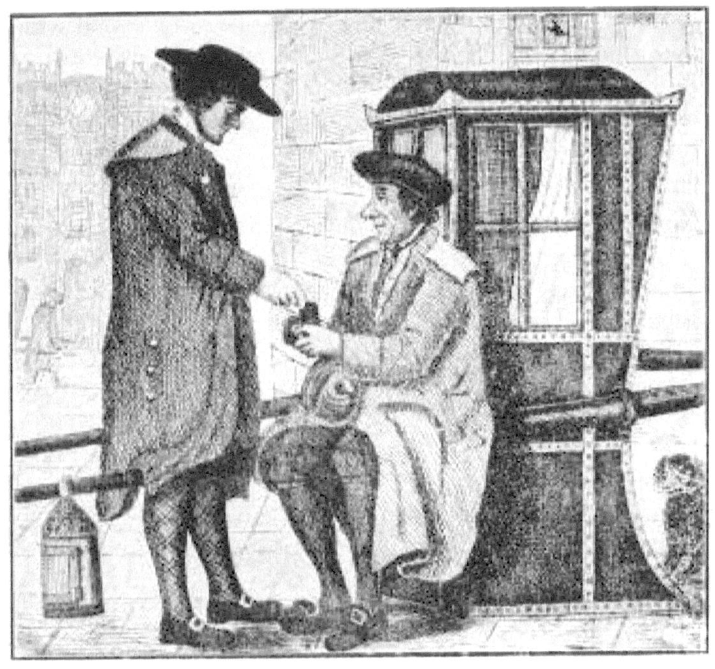

"The Social Pinch"

By John Kay

Sedans in "The Present Age"

By L. P. Boitard (1767)

In general the roof opened upwards, being hinged at the back. This is clearly shown in a print published in 1768, called *The Female Orators*, in which a clergyman is stepping out of his chair, and the chairmen very obviously demanding their fare. Another print published about 1786, called the *Social Pinch*, shows a very famous chairman, Donald Kennedy, offering his "mull" to Donald Balack, a native of Ross-shire, whom he had just set down. Here the structure of the public chair in use at this date is clearly shown.

At the beginning of the nineteenth century, however, the chair as a mode of conveyance was on the wane. Fenimore Cooper in his *Sketches of English Society* (1837) was able to write: "Sedan chairs appear to have finally disappeared from St. James' Street. Even in 1826 I saw a stand of them that has since vanished. The chairs may still be used on particular occasions, but were Cecilia now in existence, she would find it difficult to be set down in Mrs. Benfield's entry from a machine so lumbering." Which suggests that the chair had not only degenerated in numbers, but also in appearance. They had become larger and uncouth in Cooper's day. One is reminded of that chair in *Pickwick*, which "having been originally built for a gouty gentleman with funded property, would hold Mr. Pickwick and Mr. Tupman at least as comfortably as a modern post-chaise." Yet so late as 1775 the popularity of the chair had been at its highest. It was the old story. With the new century were coming new ideas. The chair slowly and quite naturally was dropping out of existence.

In Edinburgh, as I have said, it lingered on for rather a longer time. In 1806 stringent regulations were still required. Those chairs which maintained their stand at night had to have "a light fixed on the fore part of one of the poles." On the occasion of a fire or a mob the chairmen had to hurry to the scene of excitement, and there await the magistrate's orders. They were not allowed to charge more than ninepence a mile, seven-and-six a day, or a guinea and a half a week. Such rates, too, continued to be set out in the *Edinburgh Almanac* until 1830. After that comes an ominous silence. By that time only the private chair was in use.

> "Lady Don," says Lord Cockburn in his *Memorials*, "was about the last person (so far as I recollect) in Edinburgh, who kept a private sedan chair. Hers stood in the lobby and was as handsome and comfortable as silk, velvet, and gilding could make it. And when she wished to use it two well-known respectable chairmen, enveloped in her livery cloaks, were the envy of her [superannuated] brethren. She and Mrs. Rochead both sat in Tron Church; and well do I remember how I used to form one of the cluster that always took its station to see these beautiful relics emerge from coach and chair."

The time, indeed, had come when the sight of a chair was as much a public entertainment as it had been when Buckingham had been borne through the streets "on men's shoulders."

Yet although they so rapidly disappeared off the face of Europe, in Asia they lost little of their popularity, and in many places to-day are the only methods of conveyance in common use. China, in particular, had long been a land of sedans. John Barrow in his *Collection of Authentic, Useful, and Entertaining Voyages and Discoveries*, 1765, mentions the fact that at an early date the Chinese "small covered carriages on two wheels, not unlike in appearance to our funeral hearses, but only about half their length," had been superseded by chairs. To a European, he relates, this was hardly surprising, as the carriage was anything but comfortable, and required you to sit on your haunches at the bottom—"the most uneasy vehicle that can be imagined."

> "'The Chinese,' records another eighteenth-century traveller, 'occasionally travel on horseback, but their best land conveyance by far is the sedan, a vehicle which certainly exists among them in perfection. Whether viewed with regard to lightness, comfort, or any other quality associated with such mode of carriage, there is nothing so convenient elsewhere. Two bearers place upon their shoulders the poles, which are thin and elastic and in shape something like the shafts of a gig, connected near the ends, and in this manner they proceed forward with a measured step in an almost imperceptible motion, and sometimes with considerable speed. Instead of panels, the sides and back of the chair consist of woollen cloth for the sake of lightness with a covering of oil-cloth against rain. The front is closed with a hanging blind of the same materials in lieu of a door, with a circular aperture of gauze to see through.... Private persons among the Chinese are restricted to two bearers, ordinary magistrates to four, and the viceroys to eight, while the Emperor alone is great enough to require sixteen.'"

There is further mention of these Chinese chairs in Oliphant's much later account of Lord Elgin's mission. Lord Elgin himself travelled in a chair of the kind usually reserved for mandarins of the highest rank, which was larger than those in ordinary use and had a fine brass knob on the top. Eight bearers carried it. In processions a *hwakeaou* or flowered chair was often used.

Japan, too, had early had sedans both for travelling and for more purely ceremonial purposes. Light bamboo chairs, they were, called *kangoes* or *norimons*, which were borne by two or more persons. On the introduction of the European coach, however, a kind of brouette, as I have said, was substituted, and in a few years there were hundreds of thousands of these *jin-rick-shaws* on the streets, not only in Japan, but throughout Asia. At first many of these were grotesquely adorned, but their appearance is too well-known at the present day for need of a lengthy description. Equipped with "every modern convenience" and very well built indeed, they afford a European a delightful sensation on his first ride, even though he may have visions of those earlier days of his youth when he was carried about in a similar way (though at a less speed) in the homely perambulator.

CHAPTER THE FIFTH

SEVENTEENTH-CENTURY INNOVATIONS

"We took our coach, two coachmen and four horses,
And merrily from London made our courses.
We wheel'd the top of th' heavy hill called Holborne
(Up which hath been full many a sinful soule borne,)
And so along we jolted past St. Gileses,
Which place from Brainford six (or neare) seven miles is."

Taylor.

THE seventeenth century saw great changes in vehicular design. In 1660 the first *berlin* was made. Steel springs, as we have seen, appeared a few years later in the brouette. About this time, too, a hooded gig or *calèche* made its appearance in the streets of Paris, the first of many carriages to be built upon entirely new lines. Glass windows and complete doors were used in the coaches, both public and private, which became smaller, more compact, and certainly more graceful. Improvements were not confined to one country, but proceeded simultaneously not only in various European countries, but also in South America. Roads, too, were improved, and laws for the regulation of traffic framed with some regularity and effect.

John Evelyn in his Diary gives interesting glimpses of such carriages and other vehicles as he saw during his several European tours. In Brussels (1641) he was allowed the use of Sir Henry de Vic's coach and six, and travelled luxuriously in it as far as Ghent. "On the way," he notes, "I met with divers little waggons, prettily contrived, and full of peddling merchandize, drawn by mastiff dogs, harnessed completely like so many coach-horses; in some four, in others six, as in Brussels itself I had observed. In Antwerp I saw, as I remember, four dogs draw five lusty children in a chariot." When dogs were first used for the purpose of traction does not appear, but they are still to be seen in the Netherlands in a like capacity. A few days later, to continue with Evelyn's observations, he was going from Ostend to Dunkirk "by waggon ... the journey being made all on the sea sands." On his return to England, however, it is to be noticed that he rode post to Canterbury. In 1643 he was again in Paris, mentioning "the multitude of coaches passing every moment over the bridge," this being, he says, to a new spectator, "an agreeable diversion." In the following year, while standing in the garden of the Tuileries, he saw "so many coaches as one would hardly think could be maintained in the whole city, going late as it was, towards the course"—the fashionable rendezvous of the day—"the circle being capable of containing a hundred coaches to turn commodiously, and the larger of the plantations for five or six coaches abreast." The road from Paris to Orleans he describes as "excellent." Coming to Italy, he found Milan, in spite of the narrowness of its streets, abounding in rich coaches. In Paris again, two or three years afterwards, the design of a new coach so took his fancy that he determined, like his friend Mr. Pepys, to possess one for himself. And so on May 29th, 1652, "I went," he writes, "to give orders about a coach to be made against my wife's coming, being my first coach, the pattern whereof I brought out of Paris." This was probably "booted," but differed from the earlier coaches in having a curved roof.

The commonest French coach of this time seems to have been the *corbillard*, a flat-bottomed, half-open, half-close coach, furnished with curtains of cloth or leather in the front part. These were merely tied on to the supports, and would roll up when required. Doors there were none, but there was a "movable rail, over which a leather screen was hung" at the back portion of the carriage, which was about six feet long, and here were the seats. There were also projecting movable step-seats. Possibly Evelyn saw a newer model with a curved bottom and door half-way up, panelled in the lower part, but curtained above. Such a carriage was hung low, and would have swung from side to side, giving such passengers as were "bad sailors" a fit of nausea.

The English-designed coaches of this time, though without glass windows, were almost completely enclosed, and, compared with the new *chariots*, which were just upon making their appearance, of a huge size. In many of them three people could sit abreast, and seven or eight find room for themselves. In 1641 when Charles I passed through London on his return from Scotland, his was the only coach in the royal procession, but seven people, including His Majesty, were driving, apparently in comfort, within it.

The Commonwealth produced no new carriage, although isolated experiments were already being made. Cromwell himself was wont to drive his own coach and six "for recreation-sake" in Hyde Park, then as now a fashionable resort.

> "When my Lord Protector's coach," wrote Misson, a Frenchman then on a visit to England, "came into the Park with Colonel Ingleby and my Lord's three daughters, the coaches and horses flocked about them like some miracle. But they galloped (after the mode court-pace now) round and round the Park, and all that great multitude hunted them and caught them still at the turn like a hare, and then made a lane with all reverent haste for them, and so after them again, and I never saw the like in my life."

Cromwell's desire to play coachman once led to an accident which might have been serious. The particulars are given in a letter from the Dutch Ambassador to the States-General, dated October 16th, 1654:—

> "His Highness, only accompanied with secretary Thurloe and

some few of his gentlemen and servants, went to take the air in Hyde Park, when he caused some dishes of meat to be brought, when he had his dinner; and afterwards had a mind to drive the coach himself. Having put only the secretary into it," he whipped up "those six grey horses, which the Count of Oldenburgh had presented unto His Highness, who drove pretty handsomely for some time. But at last, provoking these horses too much with the whip, they grew unruly and ran so fast that the postillion could not hold them in, whereby His Highness was flung out of the coach upon the pole.... The secretary's ankle was hurt leaping out, and he keeps his chamber."

Coach in the time of Charles I

(From "Coach and Sedan Pleasantly Disputing")

"From this," comments Sir Walter Gilbey, who quotes the letter, "it is evident that when six horses were used a postillion rode one of the leaders and controlled them; while the driver managed the wheelers and middle pair. When four horses were driven," he continues, "it was the custom to have two outriders, one to ride at the leaders' heads, and one at the two wheelers'. In town this would be merely display, but on a journey the outriders' horses might replace those of the team in case of accident, or, more frequently, be added to them to help drag the coach over a stretch of bad road."

Coach in the time of Charles II

(From Thrupp's "History of Coaches")

It is just possible that this coach which was overturned by Cromwell's faulty driving is at present in existence, repaired, of course, and redecorated, and, incidentally, painted by Cipriani, as Mr. Speaker's coach. This undoubtedly belongs to the period, and one writer actually commits himself to the statement that the two are identical. A commoner report assigns the Speaker's coach in the first place to Lenthall, Cromwell's Speaker. Whatever be its history, the coach is a fine example of Jacobean work. It is of carved oak, the body being hung upon leather braces. The workmanship, Mr. Oakley Williams thinks,[28] is Flemish. Cipriani's work, added late in the eighteenth century, is still in good preservation. Five people can comfortably sit inside. "The Speaker," says Mr. Williams, "presumably occupied the seat of honour alone. Opposite him sat his Chaplain and the Sergeant-at-Arms. For the accommodation of his other attendants ... a low bench is arranged across the floor of the coach, with a semicircular space for the legs of its occupants scooped out against either door" — relic, of course, of the boot. "The coach," he continues, after mentioning that the Speaker always has his own arms painted on the side of the body, and is allowed an escort of a single Lifeguardsman, "weighs two tons one hundredweight and several pounds, yet for all its size it so beautifully hung and balanced that an able-bodied man was able without undue effort to draw it out for my inspection. Its coach-house is one of the vaults in the inner courtyard of the House of Lords." Both origin and subsequent history of this coach, however, are wrapped in an impenetrable mystery.

[28] In an article in the *Pall Mall Magazine* for March, 1912.

Cromwell's mishap naturally gave the Royalist writers an opportunity for satire. Cleveland wrote the following lines:—

> "The whip again; away! 'tis too absurd
> That thou should lash with whipcord now, but sword.
> I'm pleased to fancy how the glad compact
> Of Hackney coachmen sneer at the last act.
> Hark! how the scoffing concourse hence derives
> The proverb, 'Needs must go when th' devil drives.'
> Yonder a whisper cries, ''Tis a plain case
> He turned us out to put himself in place;
> But, God-a-mercy, horses once for aye
> Stood to 't, and turned him out as well as we.'
> Another, not behind him with his mocks,
> Cries out, 'Sir, faith, you were in the wrong box.'
> He did presume to rule because, forsooth,
> He's been a horse-commander since his youth,
> But he must know there's a difference in the reins
> Of horses fed with oats and fed with grains.
> I wonder at his frolic, for be sure
> Four hamper'd coach-horses can fling a brewer;
> But pride will have a fall; such the world's course is.
> He [who] can rule three realms can't guide four horses;
> See him that trampell'd thousands in their gore;
> Dismounted by a party but of four.
> But we have done with 't, and we may call
> The driving Jehu, Phaeton in his fall.
> I wish to God, for these three kingdoms' sake,
> His neck, and not the whip, had giv'n the crack."

Evelyn met with a similar mishap, but fortunately escaped injury. He, too, was accustomed to ride in Hyde Park, and on one occasion is grumbling that "every coach" there "was made to pay a shilling, and a horse sixpence, by a sordid fellow who had purchased it of the State, as they called it."

Such experiments as were being made in this country were in the direction of a safer and swifter vehicle than those in general use. So early as 1625, one Edward Knapp had been granted a patent for "hanging the bodies of carriages on springs of steel." Apparently Knapp was wholly unsuccessful, but forty years later Colonel Blunt, working upon similar lines, produced several carriages which, if not entirely satisfactory in themselves, led the way towards a wider appreciation of the problems in question. If, as seems probable, he was identical with the Blunt or Blount of Wicklemarsh, near Blackheath (afterwards Sir Harry Blount), who had travelled extensively in Turkey and elsewhere, it may be that he had brought back with him several continental curiosities. We hear, indeed, of a French chariot in his possession. In 1657 the Colonel was making experiments with a "way-wiser" or "adometer" which exactly "measured the miles ... showing these by an index as we went on. It had three circles, one pointing to the number of rods, another to the miles, by 10 to 1000, with all the subdivisions of quarters; very pretty," opines Evelyn, "and useful." This seems to have been the first instrument of the kind, and is overlooked by Beckmann in his account of such contrivances. The Colonel's work was brought to the notice of the newly formed Royal Society, and a committee was formed to investigate it. The first model shown to this committee was of "a chariot with four springs, esteemed by him very easy both to the rider and the horse, and at the same time cheap." The Committee also examined the designs of Dr. Robert Hooke, a distinguished member of the Society, and Professor of Geometry at Gresham College, who "produced the model of a chariot with two wheels and short double springs to be driven by one horse; the chair of it being so fixed upon two springs that the person sitting just over or rather a little behind the axle-tree was, when the experiment was made at Colonel Blunt's house, carried with as much ease as one could be in the French chariot without at all burthening the horse."[29] Dr. Hooke showed "two drafts of this model having this circumstantial difference—one of these was contrived so that the boy sitting on a seat made for him behind the chair and guiding the reins over the top of it, drives the horse. The other by placing the chair behind and the saddle on the horse's back being to be borne up by the shafts, that the boy riding on it and driving the horse should be little or no burden to the horse."

The Colonel continued experimenting both with the older coaches and a new light chariot. In 1665 Mr. Pepys was taken to see an improvement of his on a coach.

> "I met my Lord Brouncker, Sir Frederick Murrey, Dean Wilkins, and Mr. Hooke, going by coach to Colonel Blunt's to dinner.... No extraordinary dinner, nor any other entertainment good; but afterwards to the tryal of some experiments about making of coaches easy. And several we tried; but one did prove mighty easy, not here for me to describe, but the whole body of the coach lies upon one long spring, and we all, one after another, rid in it; and it is very fine and likely to take."

A few months later Pepys saw the new chariot itself.

> "After dinner comes Colonel Blunt in his new chariot made with springs; as that was of wicker, where in a while since we rode at his house. And he hath rode, he says, now his journey, many miles in it with one horse, and out-drives any coach, and out-goes any horse, and so easy he says. So for curiosity, I went into it to try it, and up the hill [Shooter's Hill] to the heath [Blackheath], and over the cart ruts, and found it pretty well, but not so easy as he pretends."

The Colonel persevered. At the beginning of the next year the Royal Society's committee met again at his house to consider, says Pepys, "of the business of chariots, and to try their new invention, which I saw here my Lord Brouncker ride in: where the coachman sits astride upon a pole over the horse, but do not touch the horse, which is a pretty odde thing; but it seems it is most easy for the horse, and,

---

[29] Birch's *History of the Royal Society.*

as they say, for the man also."

Others were also at work upon carriage improvement, and in 1667 the Royal Society "generally approved" of a chariot invented by a Dr. Croune. "No particulars of the vehicle are given," says Sir Walter Gilbey, "we are only told that 'some fence was proposed to be made for the coachman against the kicking of the horse.'" In the same year, Sir William Pen possessed a light chariot in which Pepys drove out one day. This, he says, was "plain, but pretty and more fashionable in shape than any coaches he hath, and yet do not cost him, harness and all, above £32."

All such experiments were undoubtedly in the direction of a light, swift carriage, such as was built about 1660 in Germany by Philip de Chiesa, a Piedmontese, in the service of the Duke of Prussia. Indeed, it is quite possible that Colonel Blunt either possessed, or had seen, one of de Chiesa's carriages, which were none other than the famous and popular *berlins*.[30]

So far Germany had been taking the lead. Her State coaches were the most wonderful in the world, and her coachbuilders were designing lesser coaches for the ordinary folk. But the *berlin* was the first of these lesser carriages to catch the public fancy, and enjoy more than a local success. Now the *berlin* differed in the first place from previous carriages in having two perches instead of the single pole, "and between these two perches, from the front transom to the hind axle-bed, two strong leather braces were placed, with jacks or small windlasses, to wind them up tighter if they stretched." The bottom of the coach was no longer flat, and these braces of leather allowed the body to play up and down instead of swinging from side to side as before. Here, then, you had an entirely new principle.

> "In the Imperial mews at Vienna," says Thrupp, "are four coach berlins, which, I think, may belong to this period. They are said to have been built for the Emperor Leopold who reigned at Vienna from 1658 to 1700, and Kink describes this Emperor's carriage as covered with red cloth and as having glass panels; he also says they were called the Imperial glass coaches. It is possible that the coaches have been a little altered from the time of

---

[30] Some people have considered that the name was not derived from the city of Berlin, but from an Italian word *berlina*, "a name given by the Italians to a kind of stage on which criminals are exposed to public ignominy." This seems rather far-fetched. In England it was always thought to have been built first in Berlin, and was a common enough term for a coach early in the eighteenth century. Swift mentions it in his *Answer to a Scandalous Poem* (1733):—

"And jealous Juno, ever snarling,
Is drawn by peacocks in her berlin."

"It should be noted," says Croal, "that we find the word differently applied in the earlier years of the century, and in such a way as to cast doubts on the derivations quoted. In some of the last Acts passed by the Scottish Parliaments before the Union, there are references to a kind of *ship* or boat, called a berline. The royal burghs on the west coast of Scotland were in 1705 ordered to maintain two 'berlines' to prevent the importation of 'victual' from Ireland, this importation being forbidden at the time, and two years later an Act was passed to pay the expenses of the berlines."

their construction, but I consider that in these four we have the oldest coaches with solid doors and glasses all round that exist in Europe. Whether they are identical with the Emperor Leopold's wedding-carriages matters much less than the influence the *berlin* undoubtedly had upon the coach-building of that period. It was the means of introducing the double perch, which, although it is not now in fashion, was adopted for very many carriages both in England and abroad, up to 1810. Crane-necks to perches were suggested by the form of the *berlin* perch; and as bodies swinging from standard posts suggested the position of the C spring, so bodies resting upon long leather braces suggested the horizontal and elbow springs to which we owe so much. The first *berlin* was made as a small *vis-à-vis* coach—small because it was to be used as a light travelling carriage, and narrow because it was to hang between the two perches, and was only needed to carry two persons inside. It was such an improvement in lightness and appearance upon the cumbersome coaches that carried eight persons, that it at once found favour, and was imitated in Paris and still more in London."

These early *berlins* were not nearly so gorgeous as the heavier coaches which they gradually supplanted. Red cloth and black nails had taken the place of the gilt ornamentation and crimson hangings of the previous generation.[31] Only on festivals, we learn, the black harness "was ornamented with silk fringe." The coaches used by the Emperor himself had leather traces, but the ladies of his suite had to be content with carriages the traces of which were made of rope.

The glass windows which were such a conspicuous feature of the *berlins*, were also used in the larger coaches, finally, as I have said, eliminating the boot. Mr. Charles Harper thinks that the first English coach to possess them belonged in 1661 to the Duke of York. At first these windows seem to have caused trouble, and there is the ludicrous incident mentioned by Pepys, of my Lady Peterborough who "being in her glass-coach with the glass up and seeing a lady pass by in a coach whom she would salute, the glass was so clear that she thought it had been open, and so ran her head through the glass!" Lady Ashly did not like the new invention, because, as she said, the windows were for ever flying open while the coach was running over a bad piece of road. Lady Peterborough's misfortune was tribute indeed to the maker!

In this matter of the glass it would seem that Spain had taken the lead, and it is quite possible that Spain invented the first two-seated chariots. In 1631, thirty years before the first *berlin* was made, an Infanta of Spain is reported to have traversed Carinthia "in a glass-carriage in which no more than two persons could sit." What this was like we do not know. It may have had rude springs, and been

[31] A point of minor interest may here be noticed. When leather was first used for the covering of the coach quarters, the heads of the nails showed. But about 1660, "these nail-heads were covered with a strip of metal made to imitate a row of beads; from this practice arose the name of 'beading' which has been retained, although beading is now made in a continuous, level piece, either rounded or angular." *Thrupp.*

built from the common coach models to a smaller measurement; it was certainly bootless, and framed glass or mica took the place of curtains. In France the first coaches to have glass windows, according to M. Roubo, created something of a Court scandal in the time of Louis XIII. The glass, he says, was first used in the upper panels of the doors, but was soon extended to the whole of the upper half of the sides and front of the body, so making of the carriage literally a glass-coach.

You may learn more of the English seventeenth-century carriages from Pepys than from any other writer; nor is this a matter for wonder. Pepys had a knack of knowing just exactly what posterity would desire to know. From his Diary, we learn incidentally that the watermen were still endeavouring to regain their lost prestige and custom, but by this time coaches had enormously increased in number—in 1662 there were nearly 2500 hackneys in London alone—and thenceforth they are hardly heard of. To be any one, moreover, you had to have your private coach. Doctors, for instance, found it very well worth their while to keep a coach, though, as Sir Thomas Browne told his son, they were certainly "more for state than for businesse." On the other hand those who were well able to keep a private carriage occasionally preferred the use of a hackney, and sometimes at times when they had no business to do so. Mr. Pepys, with clear ideas upon the dignity and responsibilities of rank, was indignant at any such foolery. He was told, he recalls in one place, "of the ridiculous humour of our King and Knights of the Garter the other day, who, whereas heretofore their robes were only to be worn during their ceremonies and service, these, as proud of their coats, did wear them all day till night, and then rode into the Park with them on. Nay, and he tells us he did see my Lord Oxford and Duke of Monmouth in a hackney-coach with two footmen in the Park, with their robes on; which is a most scandalous thing, so as all gravity may be said to be lost amongst us."

The private coach, too, was the last luxury to be given up after financial embarrassment. So we have Lady Flippant, in Wycherley's *Love in a Wood*, saying, "Ah, Mrs. Joyner, nothing grieves me like the putting down my coach! For the fine clothes, the fine lodgings,—let 'em go; for a lodging is as unnecessary a thing to a widow that has a coach, as a hat to a man that has a good peruke. For, as you see about town, she is most probably at home in her coach:—she eats, and drinks, and sleeps in her coach; and for her visits, she receives them in the playhouse." No lady's virtue, according to this cynical dramatist, was proof against a coach and six.

At the time of the introduction of the light, two-seated chariots, ordinary private coaches were also changing in shape. In Charles I's reign they had been both very long and very wide; in his son's time they became much slenderer and less unwieldy. Alterations in this direction were possibly suggested by the ubiquitous and most convenient sedans, and, indeed, there is an allusion to this change of shape in Sir William Davenant's *First Day's Entertainment at Rutland House*, in which, during a dialogue between a Russian and a Londoner, the foreigner says: "I have now left your houses, and am passing through your streets; but not in a coach, for they are uneasily hung, and so narrow that I took them for sedans upon wheels."

Stage-coaches, however, remained just as huge and just as gorgeous as ever.

They were built, more particularly in Italy, in the old fashion—unenclosed and curtained. Count Gozzadini describes a State coach built in 1629 for the marriage of Duke Edward Farnese with the Lady Margaret of Tuscany, and as we shall see in a moment, this differed only in the details of its ornamentation from the State coach in which Lord Castlemaine made his public entry into Rome sixty years later.

The body of the Farnese coach, says Gozzadini, "was lined with crimson velvet and gold thread, and the woodwork covered with silver plates, chased and embossed and perforated, in half relief. It could carry eight persons, four on the seats attached to the doors, and four in the back and front. The roof was supported by eight silver columns, on the roof were eight silver vases, and unicorns' heads and lilies in full relief projected from the roof and ends of the body here and there. The roof was composed of twenty sticks, converging from the edge to the centre, which was crowned with a great rose with silver leaves on the outside, and inside by the armorial bearings of the Princes of Tuscany and Farnese held up by cupids. The curtains of the sides and back of the coach were of crimson velvet, embroidered with silver lilies with gold leaves. At the back and front of the coach-carriage were statues of unicorns, surrounded by cupids and wreathed with lilies, grouped round the standards from which the body was suspended; on the tops of the standards were silver vases, with festoons of fruit, and wrought in silver. In the front were also statues of Justice and Mercy, supporting the coachman's seat. The braces suspending the body were of leather, covered with crimson velvet; the wheels and pole were plated with polished silver. The whole was drawn by six horses, with harness and trappings covered with velvet, embroidered with gold and silver thread, and with silver buckles. It is said that twenty-five excellent silversmiths worked at this coach for two years, and used up 25,000 ounces of silver; and that the work was superintended by two master coachbuilders, one from Parma and the other from Piacenza." Lord Castlemaine's procession into Rome contained three hundred and thirty coaches, of which thirteen were his own property; and of these two were State coaches. These likewise were not properly enclosed, and had no glass.

> "They were hung," says Thrupp, "inside and out, with beautifully embroidered cloths, the one coach with crimson, the other with azure-blue velvet, and gold and silver work. The roofs were adorned with scroll work and vases gilt; under the roof were curtains of silver fringes, and the ambassador's armorial bearings. The carriage of the principal coach was adorned in front with two large Tritons, of carved wood, gilt all over, that supported a cushion for the coachman between them, and from their shoulders the braces depended. The footboard was formed by a conch shell, between two dolphins. In the rear of the coach were two more Tritons, supporting not only the leather braces of the coach, but two other statues of Neptune and Cybele, who in turn held a royal crown. Below Neptune and Cybele, and projecting backwards, were a lion and a unicorn, and several cupids and wreaths

of flowers. The wheels had moulded rims, and the spokes were hidden by curving foliage carving. The second coach had plainer wheels and fewer statues about it."

They may have been magnificent, but they were certainly not very beautiful. Much the same, too, might be said of those coaches in which foreign ambassadors made their public entry into London. In 1660 Evelyn saw the Prince de Ligne, Ambassador-Extraordinary from Spain, make a splendid entry with seventeen coaches, and a month later Pepys was watching "the Duke de Soissons go from his audience with a very great deal of state: his own coach all red velvet covered with gold lace, and drawn by six barbes, and attended by twenty pages very rich in cloths."

In this year, 1660, there was a proclamation against the excessive number of hackney-coaches, and two years later Commissioners were appointed "for reforming the buildings, ways, streets and incumbrances, and regulating the hackney-coaches in the city of London." Of this body Evelyn was sworn a member in May, 1662. Pepys, however, never found any difficulty in obtaining one when he desired, and, indeed, of late years, pressure of business had made a hackney-coach an almost daily necessity. Finally, he found it cheaper to possess one of his own, and the story of this coach is particularly interesting, and may be told in some detail.

Long ago, Mr. Pepys had dreamt of owning a private coach. "Talking long in bed with my wife," he writes on March 2nd, 1661-2, "about our frugal life for the time to come, proposing to her what I could and would do, if I were worth £2000, that is, be a knight, and keep my coach, which pleased her." Times were bad, however, and although Pepys enjoyed many a ride in a friend's coach and witnessed Colonel Blunt's experiments, the great idea did not mature. But one of his particular friends, Thomas Povey, M.P., who had been a colleague of his on the Tangier committee, himself the owner of at least one coach, seems to have kept Pepys's ambitions astir. This was more especially the case in 1665, at which time Mr. Povey had purchased one of the new and already fashionable chariots. This excited Pepys's admiration. "Comes Mr. Povey's coach," he records, "and so rode most nobly, in his most pretty and best-contrived chariot in the world, with many new contrivances, his never having till now, within a day or two, been yet finished." Povey was something of an inventor himself. Evelyn calls him a "nice contriver of all elegancies, and most formal." The necessary money was apparently not forthcoming for a year or two, but in April, 1667, Pepys had a mind "to buy enough ground to build a coach-house and stable; for," says he, "I have had it much in my thoughts lately that it is not too much for me now, in degree or cost, to keep a coach, but contrarily, that I am almost ashamed to be seen in a hackney." Accordingly, Mr. Commander, his lawyer, was bidden to look for a suitable piece of ground. The idea had now taken definite shape, and Pepys was committed. "I find it necessary," he says, "for me, both in respect of honour and the profit of it also, my expence in Hackney coaches being now so great, to keep a coach, and therefore will do it." The next entry shows the first of his disappointments:—

"Mr. Commander tells me, after all, that I cannot have a lease of the ground for my coach-house and stable, till a lawsuit be ended. I am a little sorry, because I am pretty full in my mind of keeping a coach; but yet," he adds philosophically—the date was June 4th, 1667—"when I think of it again, the Dutch and French both at sea, and we poor, and still out of order, I know not yet what turns there may be."

So the summer passed, and "most of our discourse," he admits, "is about our keeping a coach the next year, which pleases my wife mightily; and if I continue as able as now, it will save me money." At the beginning of the new year Will Griffin was ordered to make fresh inquiries about the most necessary coach-house, but nothing seems to have been done until the autumn. Then Pepys, more or less it would seem on the spur of the moment, chose a coach for himself, and immediately disliked it. No one seems to have given him the same advice. Some ladies, for instance, Mrs. Pepys amongst them, preferred the large old-fashioned coaches. Others wanted the latest thing from Paris. Says Mrs. Flirt in *The Gentleman Dancing-Master*: "But take notice, I will have no little, dirty, second-hand chariot, new furnished, but a large, sociable, well-painted coach; nor will I keep it till it be as well-known as myself, and it comes to be called Flirt-coach." Her friend, Monsieur Paris, shrugs his shoulders. "'Tis very well," says he, "you must have your great, gilt, fine painted coach. I'm sure they are grown so common already amongst you that ladies of quality begin to take up with hackneys again." It was felt, no doubt, that fashion in carriages as in everything else would speedily change. Mr. Pepys must have found considerable difficulty in making up his mind. The new chariots were small, light and, so far as he knew, most fashionable; but possibly they were not quite to his taste, and equally possibly they might not be fashionable in ten years' time. Also they perhaps lacked the solid dignity of the older carriages, and were less likely to attract public attention—two important considerations. In the end, however, he seems to have chosen a large coach of the old style. Mr. Povey saw it, and poor Pepys knew at once that a dreadful mistake had been made.

"He and I ... talk of my coach," runs the Diary for 30th October, "and I got him to go and see it, where he finds most infinite fault with it, both as to being out of fashion and heavy, with so good reason, that I am mightily glad of his having corrected me in it; and so I do resolve to have one of his build, and with his advice, both in coach and horses, he being the fittest man in the world for it."

Accordingly on the following Sunday, "Mr. Povey sent his coach for my wife and I to see, which we liked mightily, and will endeavour to have him get us just such another." Mr. Povey thought that his own coachmaker had a replica for sale. Pepys thereupon went down into the neighbourhood of Lincoln's Inn Fields, found the man, but learnt to his disgust that the coach had been sold that very morning. At the end of the week, however, in company with his friend, he "spent the afternoon going up and down the coachmakers in Cow Lane, and did see several, and last did pitch upon a little chariott, whose body was framed, but not covered, at the widow's, that made Mr. Lowther's fine coach; and we are mightily pleased with it, it being light, and will be very genteel and sober; to be covered with leather, but yet will hold four. Being much satisfied with this, I carried him to White Hall. Home, where I give my wife a good account of the day's work."

Having bought the coach, it was necessary to complete the arrangements about a coach-house, and in the same week Pepys fared forth again for the purpose.

> "This afternoon I did go out towards Sir D. Gauden's, thinking to have bespoke a place for my coach and horses, when I have them, at the Victualling Office; but find the way so bad and long that I returned, and looked up and down for places elsewhere, in an inne, which I hope to get with more convenience than there."

This not proving satisfactory, Sir Richard Ford was persuaded to lend his own coach-yard. Then follow in quick succession the other entries:—

> "*28th November, 1668.*—All the morning at the Office, where, while I was sitting, one comes and tells me that my coach is come. So I was forced to go out, and to Sir Richard Ford's, where I spoke to him, and he is very willing to have it brought in, and stand there: and so I ordered it, to my great content, it being mighty pretty, only the horses do not please me, and, therefore, resolve to have better."

> "*29th November.*—This morning my coachman's clothes come home and I like the livery mightily…. Sir W. Warren … tells me, as soon as he saw my coach yesterday, he wished that the owner might not contract envy by it; but I told him it was now manifestly for my profit to keep a coach, and that, after employments like mine for eight years, it were hard if I could not be thought to be justly able to do that."[32]

> "*30th November.*—My wife after dinner, went abroad the first time in her coach, calling on Roger Pepys, and visiting Mrs. Creed, and my cozen Turner. Thus ended this month, with very good intent, but most expenseful to my purse on things of pleasure, having furnished my wife's closet and the best chamber, and a coach and horses, that ever I knew in the world; and I am put into the greatest condition of outward state that ever I was in, or hoped ever to be, or desired; and this at a time when we do daily expect great changes in this office; and by all reports we must, all of us, turn out."

> "*2nd December.*—Abroad with my wife, the first time that ever I rode in my own coach, which do make my heart rejoice, and praise God, and pray him to bless it to me and continue it."

> "*3rd December.*— … and so home, it being mighty pleasure to go alone with my poor wife, in a coach of our own, to a play,

---

[32] See below, p. 74.

and makes us appear mighty great, I think, in the world; at least, greater than ever I could, or my friends for me, have once expected; or, I think, than ever any of my family ever yet lived, in my memory, but my cozen Pepys in Salisbury Court."

"*4th December.* —I carried my wife ... to Smithfield, where they sit in the coach, while Mr. Pickering, who meets me at Smithfield and I, and W. Hewer and a friend of his, a jockey did go about to see several pairs of horses, for my coach; but it was late, and we agreed on none, but left it to another time: but here I do see instances of a piece of craft and cunning that I never dreamed of, concerning the buying and choosing of horses."

There were plenty of horses to be had, it seems, but either Mr. Pepys did not like them or he was afraid of being cheated. "Up and down," he is recording a week or so later, "all the afternoon about horses, and did see the knaveries and tricks of jockeys. At last, however, we concluded upon giving £50 for a fine pair of black horses we saw this day se'nnight; and so set Mr. Pickering down near his house, whom I am much beholden to, for his care herein, and he hath admired skill, I perceive, in this business, and so home." So the horses were changed, and for a while Mr. Pepys was obliged to revert to the despised hackney, his "coachman being this day about breaking of my horses to the coach, they having never yet drawn." Towards the end of the month the new horses were ready, and their master made his first ride behind them on a visit to the Temple, though later in the day he was again using the old pair, "not daring yet to use the others too much, but only to enter them." Then, before the new year, came the first mishap.

"Up, and vexed a little to be forced to pay 40s. for a glass of my coach, which was broke the other day, nobody knows how, within the door, while it was down; but I do doubt that I did break it myself with my knees."

At the beginning of February another misfortune is recorded: —

"Just at Holborn Circuit the bolt broke, that holds the forewheels to the perch, and so the horses went away with them, and left the coachman and us; but being near our coachmaker's and we staying in a little ironmonger's shop, we were presently supplied with another."

Accidents of this kind were continually happening. Glasses smashed, bolts broke, and, what seems incredible, doors were lost! Even so late as 1710, a reward of 30s. was offered for a lost door. "Lost," runs this remarkable advertisement, "the side door of a Chariot, painted Coffee Colour, with a Round Cipher in the Pannel, Lin'd with White Cloath embos'd with Red, having a Glass in one Frame, and White Canvas in another, with Red Strings to the Frames."

To return to Pepys. In a month or two another matter connected with his coach was occupying his attention. There were some people who did not think that a man in the comparatively humble position of Secretary to the Admiralty had any right to possess a coach, even though, in its owner's estimation, it might be "genteel and sober."

> "To the Park," he is recording in April, "my wife and I; and here Sir W. Coventry did first see me and my wife in a coach of our own; and so did also this night the Duke of York, who did eye my wife mightily. But I begin to doubt that my being so much seen in my own coach at this time, may be observed to my prejudice, but I must venture it now."

This was no idle fear, for in a while there was printed an ill-written and scurrilous pamphlet called *Plane Truth, or Closet Discorse betwixt Pepys and Hewer*, in which the following passage occurs:—

> "There is one thing more you must be mightily sorry for with all speed. Your presumption in your coach in which you daily ride as if you had been son and heir to the great Emperor Neptune, or as if you had been infallibly to have succeeded him in his government of the Ocean, all which was presumption in the highest degree. First, you had upon the fore-part of your chariot, tempestuous waves and wrecks of ships; on your left hand, forts and great guns, and ships a fighting; on your right hand was a fair harbour and galleys riding, with their flags and pennants spread, kindly saluting each other, just like P[epys] and H[ewer—his chief clerk]."

How far Pepys's carriage was decorated is not known, though this description does not tally in the least with Pepys's own. In any case, he took no notice of such attacks, and so far from making his coach less conspicuous, arranged to have it newly painted and varnished.

> "*19th April, 1669.*—After dinner out again, and, calling about my coach, which was at the coachmaker's, and hath been there for these two or three days, to be new painted, and the window-frames gilt against next May-day, went on with my hackney to White Hall."

A few days later he gave orders for some "new sort of varnish" to be used on the standards at a cost of forty shillings, this being in his view very cheap. Indeed, "the doing of the biggest coach all over," he learnt, "comes not above £6." On his next visit to the coachmaker, he was surprised to find several great ladies "sitting in the body of a coach that must be ended tomorrow ... eating of bread and butter and drinking ale." His own coach had been silvered over, "but no varnish yet laid on, so I put it in a way of doing." A few hours later he called back again,

> "and there vexed to see nothing yet done to my coach, at three in the afternoon; but I set it in doing, and stood by till eight at night, and saw the painter varnish it which is pretty to see how every doing it over do make it more and more yellow: and it dries as fast in the sun as it can be laid on almost; and most coaches are, now-a-days, done so, and it is very pretty when laid on well, and not too pale, as some are, even to show the silver. Here I did make the

workmen drink, and saw my coach cleaned and oyled."

And so eager was he to have it without delay that his coachman and horses were sent to fetch it that very evening, and on the following gala day, May 1st,

> "we went alone through the town with our new liveries of serge, and the horses' manes and tails tied with red ribbons, and the standards gilt with varnish, and all clean, and green reines, the people did mightily look upon us; and, the truth is, I did not see any coach more pretty, though more gay, than ours all the day. But we set out, out of humour—I because Betty, whom I expected, was not come to go with us; and my wife that I would sit on the same seat with her, which she likes not, being so fine: and she then expected me to meet Sheres, which we did in Pell Mell, and against my will, I was forced to take him into the coach, but was sullen all day almost, and little complaisant; the day being unpleasing, though the Park full of coaches, but dusty and windy, and cold, and now and then a little dribbling of rain; and what made it worse, there were so many hackney-coaches as spoiled the sight of the gentlemen's; and so we had little pleasure."

Henceforth Mr. Pepys, in spite of sundry warnings from his friend Mr. Povey and others, continued to use his coach, and although perhaps as he grew older, his coach was less brilliantly adorned, there seems no reason to suppose that he ever regretted its purchase.

Though it is not my intention to speak in any detail of public conveyances, a word must be said here of the stage-coaches,[33] which made their appearance on English roads in 1640. These were large coaches, leather-curtained at first—glass does not seem to have been used until 1680—and capable of seating six or eight passengers. Their chief feature was the huge basket strapped to the back.

> "There is of late," says Chamberlayne in his well-known *Present State of Great Britain* (1649), "such an admirable commodiousness both for men and women, to travel from London to the principal towns in the country, that the like hath not been known in the world; and that is by stage-coaches, wherein one may be transported to any place sheltered from foul weather and foul ways, free from endangering of one's health and one's body by hard jogging or over-violent motion on horseback; and this not only at the low price of about a shilling for every five miles, but with such velocity and speed in an hour as the foreign post can make but in one day."

Of course, there was opposition to these public coaches. In 1662, when there was not a round dozen of them, one writer was already exhorting their extinction on the ground that simple country gentlemen and their simple country wives

---

[33] The reader is referred for the fullest information on the subject of these stage-coaches to Mr. Charles G. Harper's *Stage-Coach and Mail in Days of Yore*. 2 vols. London, 1903.

could now come to London without due occasion, and there learn all the vice and luxury that were rampant. So in 1673, in a singular production called *The Grand Concern of England*, amongst the many proposals set forth for the country's good, was one "that the Multitude of Stage Coaches and Caravans be suppressed." One or two pamphlets of no particular interest appeared, both for and against these coaches, but it may be sufficient here to observe that they steadily increased in numbers and maintained their existence until the mail-coaches finally superseded them.

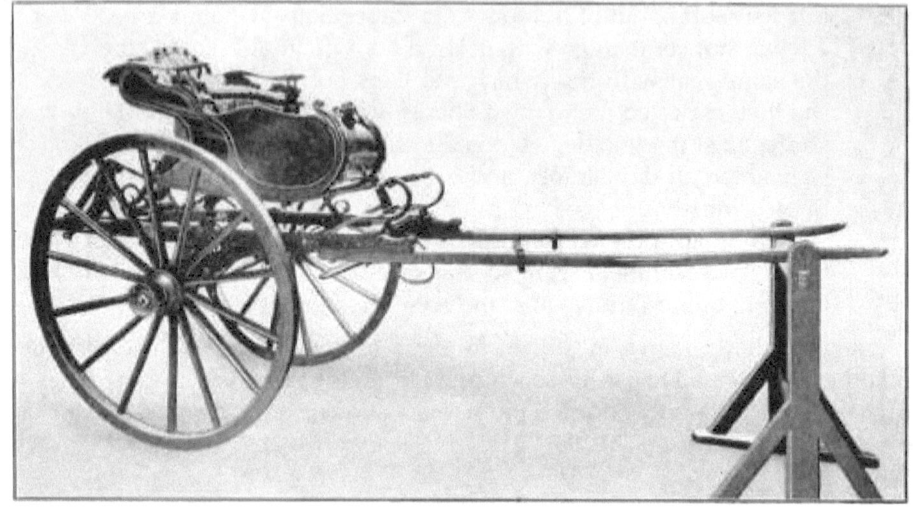

Early (?) French Gig

at the South Kensington Museum

One other public carriage of this time also deserves mention. This was the *carosse à cinq sous*, which appeared in the streets of Paris in 1662. The history of this primitive omnibus is well told by Mr. Henry Charles Moore.[34]

> "The leading spirits in this enterprise were the Duc de Rouanès, Governor of Poitou, the Marquis de Sourches, Grand Prévôt, the Marquis de Crenan, Grand Cup-bearer, and Blaise Pascal, the author of *Lettres Provinciales*. The idea was Pascal's, but not being sufficiently wealthy to carry it out unaided, he laid the matter before his friend the Duc de Rouanès, who suggested that a company should be formed to start the vehicles. Pascal consented to this being done, and the Duc set to work at once to prevail upon members of the aristocracy to take shares in the venture."
> After obtaining a royal decree, "seven vehicles to carry eight passengers each, all inside, were built, and on March 18th, 1662, they began running. The first one was timed to start at seven o'clock in the morning, but an hour or two earlier a huge crowd had assem-

[34] *Omnibuses and Cabs.* London, 1902.

bled to witness the inauguration ceremony, which was performed by two Commissaires of the Châtelet, attired in their official robes. Accompanying them were four guards of the Grand Prévôt, twenty men of the City Archers, and a troop of cavalry. The procession, on arriving at the line of route, divided into two parts, one Commissaire and half of the attendants proceeded to the Luxembourg, and the others to the Porte St. Antoine. At the latter place three of the twopenny-halfpenny coaches were stationed, the other four being at the Luxembourg. Each Commissaire then made a speech, in which he pointed out the boon that *carosses à cinq sous* would be to the public, and laid great stress on the fact that they would start punctually at certain times whether full or empty. Moreover, he warned the people that the king was determined to punish severely any person who interfered with the coaches, their drivers, conductors, or passengers. The public was also warned that any person starting similar vehicles without permission would be fined 3000 francs, and his horses and coaches confiscated.

"At the conclusion of his address, the Commissaire commanded the coachmen to advance, and, after giving them a few words of advice and caution, presented each one with a long blue coat, with the City arms embroidered on the front in brilliant colours. Having donned their livery, the drivers returned to their vehicles and climbed up to their seats. Then the command to start was given, and the two vehicles drove off amidst a scene of tremendous enthusiasm. The first coach each way carried no passengers—a very unbusinesslike arrangement—the conductor sitting inside in solitary state. But the next two, which were sent off a quarter of an hour after the first, started work in earnest, and it need scarcely be said that there were no lack of passengers. The difficulty experienced was in preventing people from crowding in after the eight seats were occupied. At the beginning of every journey the struggle to get into the coach was repeated, and many charming costumes were ruined in the crush. Paris, in short, went mad over its *carosses à cinq sous*, and the excitement soon spread to the suburbs, sending their inhabitants flocking to the city to see the new vehicles. But very few of the visitors managed to obtain a ride, for day by day the rush for seats became greater. The king himself had a ride in one coach, and the aristocracy and wealthy classes hastened to follow his example, struggling with their poorer brethren to obtain a seat. Many persons who possessed private coaches daily drove to the starting-point, and yet failed to get a drive in one for a week or two.

"Four other routes were opened in less than four months, but at last the fashionable craze came to an end, and as soon as the upper classes ceased to patronise the new coaches the middle and lower classes found that it was cheaper to walk than to ride. The result was that Pascal, who died only five months after the coaches began running, lived long enough to see the vehicles travelling to and fro, half, and sometimes quite, empty.

"For many months after Pascal's death the coaches lingered on, but every week found them less patronised, and eventually they were discontinued. They had never been of any real utility, and were regarded by the public much in the same light as we regard a switchback railway."

And, indeed, it was a century and a half before the next omnibus was tried.

So then, at the middle of the century, when heavy and slow stage-coaches were making their appearance on the English country roads, and the unsuccessful *carosse à cinq sous* was being tried in the streets of Paris, the success of the *berlin*, the *brouette*, and other chariots, was in process of remodelling men's ideas upon the most feasible carriage for town use. The older coaches, as I have said, were still retained for particular occasions, and, indeed, continued to be built with more ornamentation than ever before. The very spokes of the wheels were decorated, paintings appeared on the panels, and every inch of the coach made as brilliant as possible. France in particular possessed carriages of the most gorgeous possible description. These were not only entirely gilded over, but in some cases actually bejewelled. The richest stuffs lined their interiors, and masters painted their panels. Immense sums were spent. There is preserved at Toulouse a carriage of this date which shows most of these features. The interior "is, or rather was, lined with white brocade embroidered with a diaper of pink roses, the roof being lined with the same, while its angles are hidden by little smiling cupids gilded from top to toe. The surface of the panels is, or rather was, a piece of opaque white, exceedingly well varnished, and edged with a thick moulding of pink roses; the foliage, instead of being green, was highly gilded and burnished."

But the ever-increasing traffic rendered necessary a much smaller vehicle than these monstrosities for general use, and this led, somewhere about 1670, to the introduction of the *gig*. This was a French invention, which, while no doubt the logical outcome of the *brouette*, bore resemblance to the old Roman *cisium*, and led ultimately to the cabriolets, once so popular both in France and England. Certain experiments tending towards a gig had been made earlier in the century with a chair fixed to a small cart. The first successful gig was a slender, two-wheeled contrivance, "the body little more than a shell," says Thrupp, provided with a hood "composed of three iron hoop-sticks joined in the middle to fall upwards." It was the prototype of the *calèche* in France, the *carriole* of Norway, the *calesso* of Naples, and the *volante* of Cuba. Gozzadini describes one of them as "an affair with a curved seat fixed on two long bending shafts, placed in front on the back of the horse and behind upon the two wheels." They were introduced into Florence, he says, in 1672, and "so increased in numbers that in a few years there were nearly a thousand in the city." An early gig of this kind is preserved at South Kensington.

It is a forlorn-looking vehicle. The body is curved, but there is no hood. The seat is absurdly small and "beneath the shafts are two long straps of leather and a windlass to tighten them—this apparatus was, no doubt, to regulate the spring of the vehicle to the road travelled over."

The gig speedily underwent several minor changes of form. In France it was known as *calèche*[35] or *chaise*, in England, as *calash*, *calesh*, or *chaise*, in America as *shay*. Unfortunately there is small mention of them in contemporary writings, and one is left to suppose that for some time they did not, except in certain cities, prove serious rivals to the *berlins* and other four-wheeled chariots. It may be that the *berlin* itself was taken as a model from which these lighter carriages were evolved. You had first the big double *berlins* for four people, then you had a *vis-à-vis* for two or more persons facing each other. Later the front part of the carriage would be cut away for the sake of lightness. When not covered such a vehicle as this seems to have been known as a *berlingot*. Two could travel in these *berlingots* sitting side by side, "while a third person might travel uncomfortably in front on a kind of movable seat, which was not much patronised; for it was not only dangerous, but what was much worse in the eyes of the grand court gentlemen who used them—ridiculous." There was also evolved a smaller and narrower *berlin* with the front cut away and capable of holding only one passenger, called the *désobligeante*. The bodies of the ordinary chaises, which seated one or two people, seem to have differed from those of the older *berlins* in being placed partly below the frame. There were no side doors, but one at the back which opened horizontally. When and where all such changes were made, however, it is impossible to say. The accounts, such as they are, are often contradictory, and the same names used to describe what are obviously not identical carriages. But the two-wheeled gig having appeared there was nothing to prevent improvements of every conceivable sort or shape, and innumerable hybrid carriages appeared, some of which are only known by name.

There is mention of a truly remarkable calash which was tried in Dublin in 1685. Exactly who the inventor was is not known, but Sir Richard Bulkeley interested himself in the experiments, and read a paper on his carriage before the Royal Society. Evelyn was one of those who were present on this occasion.

> "Sir Richard Bulkeley," he says, "described to us a model of a chariot he had invented which it was not possible to overthrow in whatever uneven way it was drawn, giving us a wonderful relation of what it had performed in that kind, for ease, expedition, and safety; there were some inconveniences yet to be remedied—it would not contain more than one person; was ready to take fire every few miles; and being placed and playing on no fewer than

[35] It was over a *calèche* presented by the Chevalier de Grammont to Charles II, that the famous quarrel took place between Lady Castlemaine and Miss Stewart, afterwards the Duchess of Richmond. The ladies had been complaining that coaches with glass windows, but lately introduced, did not allow a sufficiently free display of their charms, whence followed the gift of a French *calèche* which cost two thousand livres. When the queen drove out in it, both the ladies agreed with de Grammont that it afforded far better opportunities than a coach for showing off their figures, and both endeavoured to get the first loan of it. In the fierce quarrel that followed Miss Stewart came off the conqueror.

ten rollers, it made a most prodigious noise, almost intolerable."

It is to be deeply regretted that there is no print of this remarkable carriage, but further details may be found in a letter, dated May 5th, 1685, from Sir Richard Bulkeley himself.

"Sir William Petty," he writes, "Mr. Molyneux, and I have spent this day in making experiments with a new invented calesh, along with the inventor thereof; 'tis he that was in London when I was there, but he never made any of these caleshes there, for his invention is much improv'd since he came from thence: it is in all points different from any machine I have ever seen: it goes on two wheels, carries one person, and is light enough. As for its performance, though it hangs not on braces, yet it is easier than the common coach, both in the highway, in ploughed fields, cross the ridges, directly and obliquely. A common coach will overturn, if one wheel go on a superficies a foot and a half higher than that of the other; but this will admit of the difference of three feet and a half in height of the superficies, without danger of overturning. We chose all the irregular banks, the sides of ditches to run over; and I have this day seen it, at five several times, turn over and over; that is, the wheels so overturned as that their spokes laid parallel to the horizon, so that one wheel laid flat over the head of him that rode in the Calesh, and the other wheel flat under him; so much I all but once overturned. But what I have mentioned was another turn more, so that the wheels were again *in statu quo*, and the horse not in the least disordered: if it should be unruly, with the help of one pin, you disengage him from the Calesh without any inconvenience. I myself was once overturned, and knew it not, till I looked up, and saw the wheel flat over my head; and, if a man went with his eyes shut, he would imagine himself in the most smooth way, though, at the same time, there were three feet difference in the heights of the ground of each wheel. In fine, we have made so many, and so various experiments, and are so well satisfied of the usefulness of the invention, that we each of us have bespoke one; they are not (plain) above six or eight pounds a-piece."

Why the nobility, gentry, and worthy burgesses of England, Scotland, and Ireland did not go and do likewise, history hides from us. There is no further mention of Sir Richard's truly remarkable carriage, and one is left to imagine that some of the Irish roads were too bad even for its freakish agility.

On the other hand, they were probably superior to the Scottish roads of the time, even those in the more civilised southern districts. "It is recorded," says Croal, "that in 1678"—the year after the founding of the Coach and Coach-Harness Makers' Company in London—"the difficulties in the way of rapid communication were such that an agreement was made to run a coach between Edinburgh and Glasgow, a distance of forty-four miles, which was to be drawn by six horses, and to perform the journey to Glasgow and back in six days!"

Early Italian Gig

at the South Kensington Museum

Cross-country travelling, indeed, was very bad, and the rough tracks over which the heavy stage-coaches rumbled along would have proved too much for the lighter chariots and gigs which were so popular in town. I may conclude this chapter by quoting an amusing description of such cross-country travelling at the end of the century, taken from Sir John Vanbrugh's *Provoked Husband*. A family is going in its private coach from Yorkshire to London:—

Lord Townley. Mr. Moody, your servant; I am glad to see you in London. I hope all the family is well.

John Moody. Thanks be praised, your honour, they are all in pretty good heart, thof' we have had a power of crosses upo' the road.

Lady Grace. I hope my Lady has no hurt, Mr. Moody.

John. Noa, an't please your ladyship, she was never in better humour: There's money enough stirring now.

Manly. What has been the matter, John?

John. Why, we came up in such a hurry, you mun think that our tackle was not so tight as it should be.

Manly. Come, tell us all: pray how do they travel?

John. Why i' the auld coach, Measter; and cause my Lady loves to do things handsome, to be sure, she would have a couple of cart horses clapt to th' four old geldings, that neighbours might see she went up to London in her coach and six! And so Giles Joulter the ploughman rides postilion!

*   *   *   *   *

Lord Townley. And when do you expect them here, John?

John. Why, we were in hopes to ha' come yesterday, an' it had no' been that th' owld wheaze-belly horse tired; and then we were so cruelly loaden, that the two fore-wheels came crash down at once in Waggon-Rut Lane; and there we lost four hours 'fore we could set things to rights again.

Manly. So they bring all their baggage with the coach then?

John. Ay, ay, and good store on't there is. Why, my Lady's gear alone were as much as filled four portmantel trunks, besides the great deal box that heavy Ralph and the monkey sit on behind.

Lady Grace. Well, Mr. Moody, and pray how many are there within the coach?

John. Why, there's my Lady and his Worship, and the young squoire, and Miss Jenny, and the fat lap-dog, and my lady's maid Mrs. Handy, and Doll Tripe the cook; that's all. Only Doll puked a little with riding backward, so they hoisted her into the coach-box, and then her stomach was easy.

Lady Grace. Oh! I see 'em go by me. Ah! ha!

John. Then, you mun think, Measter, there was some stowage for the belly, as well as th' back too; such cargoes of plum cake, and baskets of tongues, and biscuits and cheese, and cold boiled beef, and then in case of sickness, bottles of cherry-brandy, plague-water, sack, tent, and strong beer, so plenty as made the owld coach crack again! Mercy upon 'em! and send 'em all well to town, I say.

Manly. Ay! and well on't again, John.

John. Ods bud! Measter, you're a wise mon; and for that matter, so am I. Whoam's whoam, I say; I'm sure we got but little good e'er we turned our backs on't. Nothing but mischief! Some devil's trick or other plagued us, aw th' day lung. Crack goes one thing: Bawnce goes another. Woa, says Roger. Then souse! we are all set fast in a sleugh. Whaw! cries Miss; scream go the maids; and bawl! just as thof' they were struck! And so, mercy on us! this was the trade from morning to night.

CHAPTER THE SIXTH
EARLY GEORGIAN CARRIAGES

"May the proud chariot never be my fate,
If purchased at so mean, so dear a rate.
Oh, rather give me sweet content on foot,
Wrapt in my virtue and a good surtout."

GAY's *Trivia.*

FEW new private carriages seem to have been designed during the earlier decades of the eighteenth century, although improvements and small alterations were constantly being carried out. There is an isolated reference to a *sociable* built apparently in Germany, and the *four-wheeled chaise*, or *chariot à l'Anglaise*, which was to be so popular thirty or forty years later, put in an appearance about this time. Of the sociable little enough can be said. The particular carriage mentioned from its small size would appear to have been built for the royal children. It was a low-hung, open carriage over a single perch, and with seats facing each other. The four-wheeled chaise was a small chariot with a wide window in front.

Gray, writing to his mother in 1739, speaks of the French chaise in which he was making the grand tour with Horace Walpole.

> "The chaise," he writes, "is a strange sort of conveyance, of much greater use than beauty; resembling an ill-shaped chariot, only with the door opening before instead of the side. Three horses draw it, one between the shafts, and the other two on each side, on one of which the postillion rides, and drives too: This vehicle will upon occasion, go fourscore miles a day, but Mr. Walpole, being in no hurry, chooses to make easy journies of it, and they are easy ones indeed; for the motion is much like that of a sedan, we go about six miles an hour, and commonly change horses at the end of it. It is true they are not very graceful steeds, but they go well, and through roads which they say are bad for France, but to me they seem gravel walks and bowling-greens; in short, it would be the finest travelling in the world, were it not for the inns."

Such a chaise as Gray describes came to be known as a *diligence*, while in England the one-horse chaise was more frequently spoken of as a *one-horse chair*.

Contemporary prints of carriages, however, are scarce, and for the most part show only the larger coaches.

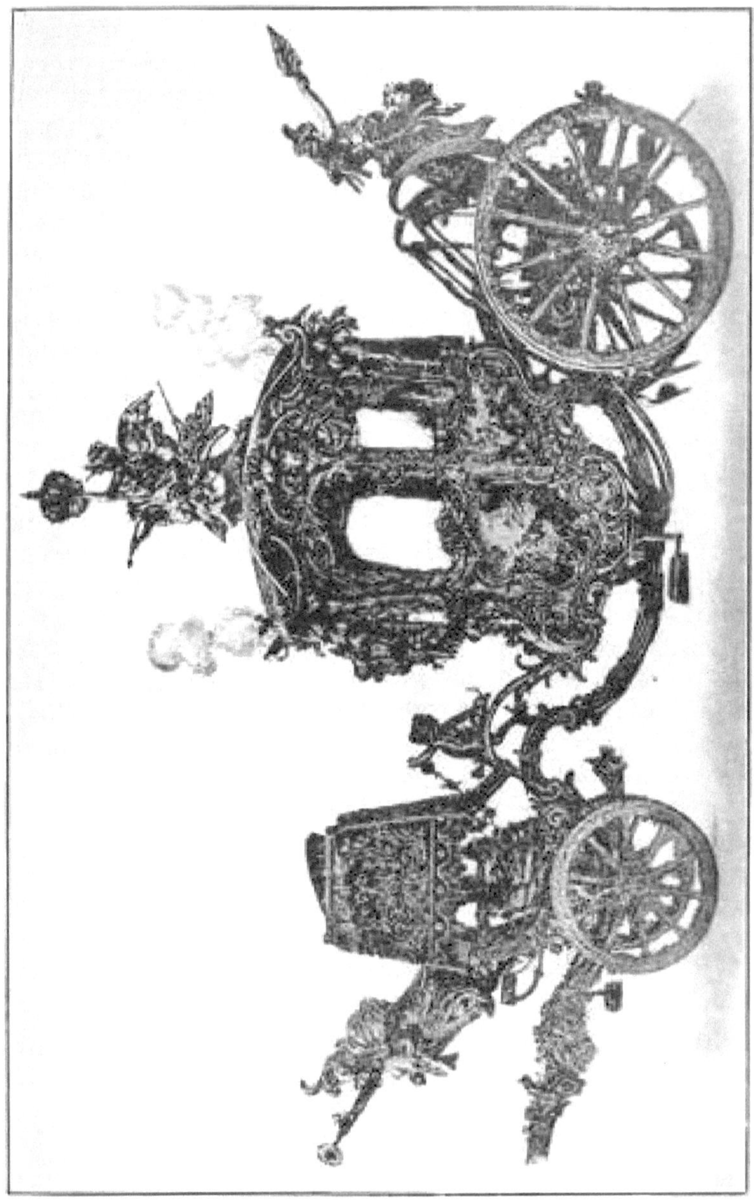

The State Carriage of Bavaria. Early Eighteenth Century

(From Smith's "Concise History of English Carriages")

These coaches were of two distinct patterns. There were the large square coaches of Charles II's time, but there was also a new type of coach or chariot which had a curious backward tilt to the body. From a superficial examination of such a carriage, it would appear impossible for the seats to have been horizontal, and, indeed, one wonders why this form was adopted. The result of this backward tilt was to leave a space between the coachman's box and the carriage-body itself. Here one of the grooms sat or sprawled as best he could. Four, five, or even six other grooms stood uncomfortably huddled together on a seat or slab at the back. These men must have added considerably to the weight of the coach, and certainly did not make travelling any swifter; but how necessary they were is shown by a letter of the period in which one nobleman's servant in London informs another in Essex that my lord is resolved to set out. The Essex man is bidden to have "the keepers and persons who know the holes and the sloughs" ready to meet his lordship "with lanterns and long poles" to keep the coach on its way. So many accidents happened even on the shortest journeys that five or six men were necessary to put the coach aright. A road, such as we think of one now, simply did not exist. You had often to drive across fields in tracks which exceedingly heavy waggons had made. In 1703, to take another instance, the King of Spain, then in this country, was journeying from Portsmouth to Windsor. The difficulties he experienced on that occasion were recorded by one of the attendants.

> "We set out at six in the morning to go to Petworth, and did not get out of the coaches (save only when we were overturned or stuck fast in the mire) till we arrived at our journey's end. 'Twas hard service for the prince to sit fourteen hours in the coach that day without eating anything and passing through the worst ways that I ever saw in my life; we were thrown but once indeed in going, but both our coach, which was the leading, and his highnesse's body-coach would have suffered very often if the nimble boors of Sussex had not frequently poised it or supported it with their shoulders from Goldalmin almost to Petworth; and the nearer we approached to the Duke's house the more unaccessible it seemed to be. The last nine miles of the way cost us six hours' time to conquer them, and indeed we had never done it if our good master had not several times lent us a pair of horses out of his own coach, whereby we were enabled to trace out the way for him."

After reading such an account, it is difficult to understand why any one preferred coach to horseback on a cross-country journey. No wonder Gay was goaded to ask:—

> "Who can recount the coach's various harms,
> The legs disjointed, and the broken arms?"

"In the wide gulph," he says in another place,

> "the shatter'd coach o'erthrown
> Sinks with the snorting steeds; the reins are broke,
> And from the crackling axle flies the spoke."

Yet, according to Swift, Gay was not so averse to the coach in his later years. Writing to him in 1731, the Dean says:—

> "If your ramble was on horseback, I am glad of it on account of your health; but I know your arts of patching up a journey between stage-coaches and friends' coaches: for you are as arrant a cockney as any hosier in Cheapside.... You love twelve-penny coaches too well, without considering that the interest of a whole thousand pounds brings you but half a crown a day."

"A coach and six horses," he goes on to say in another letter, "is the utmost exercise you can bear, and this only when you can fill it with such company as is best suited to avoid your taste, and how glad would you be if it could waft you in the air to avoid jolting."

There is preserved a chariot of this period which is probably typical of a nobleman's carriage of the time. It was built for one of the Bligh family, possibly the first Lord Darnley, about 1720. It is a small carriage, curved curiously in a fashion which recalls some of the French furniture of the period. The body is slung upon leather braces, there is a single wide perch, and there are small elbow springs under the body at the back. It is very elaborately ornamented, and still keeps some of its pristine magnificence. A curious point about the Darnley chariot, to which some people have wrongfully ascribed a much earlier date, is the length of the door, which reaches nearly a foot below the bottom of the body. A similar peculiarity is to be seen in another coach of the period which was built in 1713 for the Spanish representative at the time of the Peace of Utrecht. Here "the quarters rake towards the roof considerably, the roof over the doorway is arched upwards, the upper quarters are filled with large glasses of mirror plate glass.... The wheels have carved spokes and felloes.... There is a hammercloth cushion in front and a footboard supported by Tritons blowing horns." Another Spanish coach, with spiral spokes and similar peculiarities, is preserved at Madrid. This elongated door seems peculiar to the period and may have followed upon a desire to hide the steps, though the lowness of the carriage made more than one or two of these unnecessary. Many of the Spanish coaches of this time, by the way, were without the coach-box, postilions only being employed—the story being that a certain Duke of Olivarez found that his coachman had heard and betrayed a State secret. There was, I believe, actually a law passed in Spain forbidding coachmen altogether.

French coaches were very resplendent. "When I was in France," writes Addison in one of the earlier *Spectators*, "I used to gaze with great Astonishment at the Splendid Equipages and Party-Coloured Habits, of that Fantastick Nation. I was one Day in particular contemplating a Lady, that sate in a Coach adorned with gilded Cupids, and finely painted with the Loves of *Venus* and *Adonis*. The Coach was drawn by six milk-white Horses, and loaden behind with the same Number of powder'd Footmen. Just before the Lady were a Couple of beautiful Pages that were stuck among the Harness, and, by their gay Dresses and smiling Features, looked like the elder Brothers of the little Boys that were carved and painted in every corner of the Coach." The boys "stuck among the harness" obviously were resting in that space which was made by the back-tilting of the body.

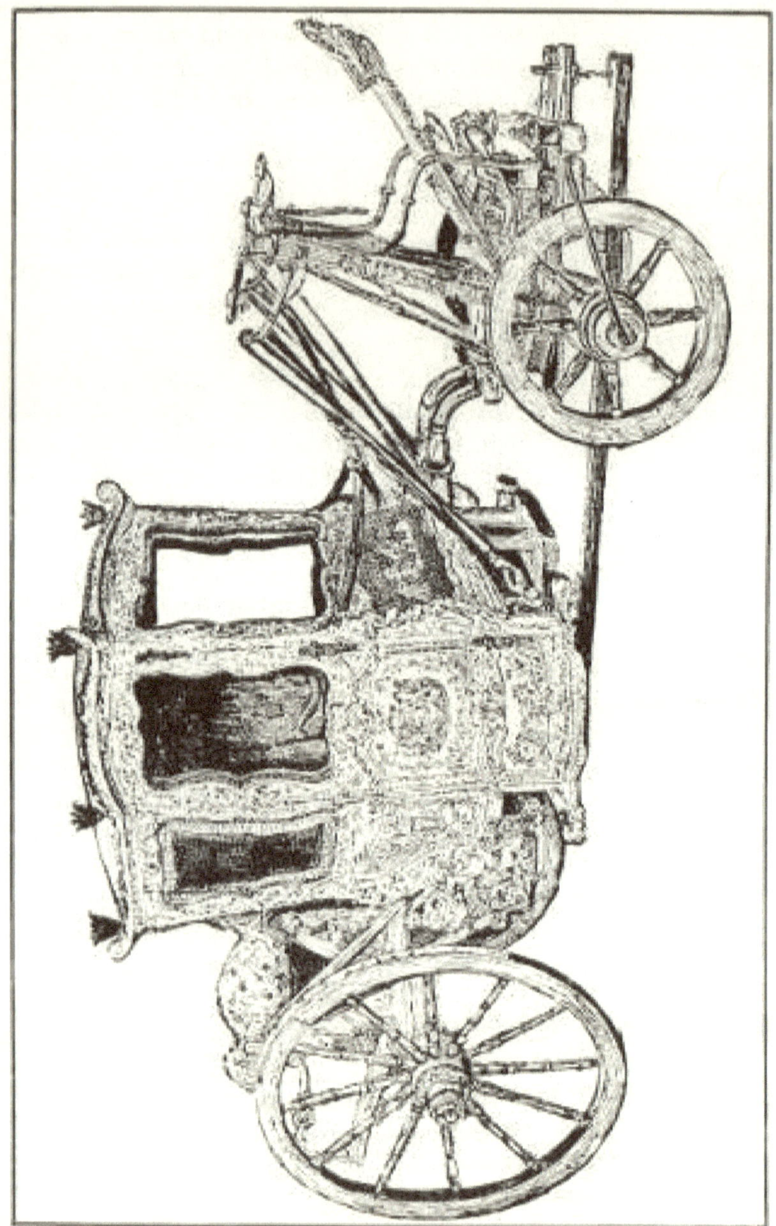

The Darnley Chariot. Early Eighteenth Century

(From Smith's "Concise History of English Carriages")

The Viennese coaches of this time seem to have had a very great deal of glass about them, but the Turkish coaches had none. Writing home from Adrianople in 1717, Lady Mary Wortley Montagu says:—

"Designing to go [to Sophia] *incognita*, I hired a Turkish coach. These voitures are not at all like ours, but much more convenient for the country, the heat being so great that glasses would be very troublesome. They are made a good deal in the manner of the Dutch coaches, having wooden lattices painted and gilded; the inside being painted with baskets and nosegays of flowers, intermixed commonly with little poetical mottoes. They are covered all over with scarlet cloth, lined with silk, and very often richly embroidered and fringed. This covering entirely hides the persons in them, but may be thrown back at pleasure, and the ladies peep through the lattices. They hold four people very conveniently, seated on cushions, but not raised."

They were, it would seem, mere covered waggons, and, indeed, in another place Lady Mary speaks of them as such. Turkey possessed also "open gilded chariots," but in these the women were not allowed to drive.

Russia, too, at this time possessed coaches, and we read that Peter the Great in his trans-European journey travelled with "thirty-two four-horse carriages and four six-horse waggons." One or two particulars are forthcoming of the royal coach-house. It contained but two coaches, with four places in each, for the use of the Empress and a smaller, low-hung carriage, painted red, for the Emperor. This was replaced in winter by a small sledge. Peter, however, was not fond of his carriage. "He never," says Waliszewski,[36] "got into a coach, unless he was called upon to do honour to some distinguished guest, and then he always made use of Menshikof's carriages. These were magnificent. Even when the favourite went out alone, he drove in a gilded fan-shaped coach, drawn by six horses, in crimson velvet trappings, with gold and silver ornaments; his arms crowned with a prince's coronet, adorned the panels; lacqueys and running footmen in rich liveries ran before it; pages and musicians, dressed in velvet, and covered with gold embroideries, followed it. Six gentlemen attended it at each door, and an escort of dragoons completed the procession."

It is difficult to conceive the appearance of this fan-shaped coach, but it must have been almost startlingly magnificent, just the kind of carriage for the Russian Buckingham.

In the imperial collection at Petersburg are preserved one or two Russian carriages of this period. "One," says Bridges Adams, "is close, made of deal, stained black, mounted on four wheels, the windows of mica instead of glass, and the frames of common tin: the other is open, with a small machine behind of the shipwright-emperor's invention—its purpose to determine the number of miles traversed on a journey. In the same collection," he adds, "is the litter of Charles XII used at the battle of Pultowa."

In England glass seems to have been reserved for the private coaches. For the commoner hackneys a substitute had been found. "For want of Glasses to our Coach," wrote the inimitable Ned Ward in *The London Spy*, a book whose outspo-

---

[36] *Peter the Great.* By K. Waliszewski. Translated by Lady Mary Loyd. London, 1898.

kenness unfortunately must, I suppose, have prevented its reprinting in modern days, "we drew up our Tin Sashes, pink'd like the bottom of a Cullender, that the Air might pass thro' the holes, and defend us from Stifling."

If, however, contemporary plates are singularly scarce, and the historians have little to say of the period, there is a new source of information to be tapped, at any rate in this country, in the advertisements which just now began to fill whole pages in the periodicals. Of these I may quote one or two. One deals specifically with the question of glass windows:—

> "These are to give notice to all Persons that have occasions for Coach Glasses, or Glasses for Sash Windows, that they may be furnished with all sorts, at half the prices they were formerly sold for."

Twelve inches square cost half a crown, thirty-six inches two pounds ten shillings.

Other advertisements concern the coaches themselves. In Anne's day calashes, chaizes, both two-and four-wheeled, as well as the larger chariots—these often flamboyantly decorated—were constantly for sale.

> "A very fine CHAIZE," we read, "very well Carved, gilded and painted, and lined with Blue Velvet, and a very good HORSE for it, are to be sold together, or apart."

> "A curious 4-Wheel SHAZE, Crane Neck'd, little the worse for wearing, it is to be used with 1 or 2 Horses, and there is a fine Harness for one Horse, and a Reputable Sumpture Laopard Covering."

Here then is mention of a four-wheeled chaise with a perch curved in front after the German fashion. Other chaises for sale had only two wheels:—

> "At the Greyhound in West Smithfield is to be sold a Two-Wheeled Chaize, with a pair of Horses well match'd: It has run over a Bank and a Ditch 5 Foot High; and likewise through a deep Pit within the Ring at Hide Park, in the presence of several persons of Quality; which are very satisfied it cannot be overturn'd with fair Driving. It is to be Lett for 7s. 6d. a Day, with some Abatement for a longer Time."

One is reminded of Sir Richard Bulkeley's wonderful calash. Here was surely a rival. Calashes were now common, though precisely what the difference was between them and the two-wheel chaises I am unable to say. Indeed, there is some confusion also between the small chariots and the four-wheel chaises, and the words seem to have become interchangeable. Both came to resemble the coupé of a later day, being like a modern coach with the front part removed. Sometimes the coachman's box was on a level with the roof, but often much lower, and sometimes altogether absent, the horses being ridden by a postilion. Probably the carriage was called a chariot when it possessed a coachman's box, such as was used in town, and a chaise when it was absent.

It was a calash that Squire Morley of Halstead wished for, but did not obtain, in Prior's ballad of *Down-Hall*, 1715.

> "Then answer'd Squire Morley; Pray get a calash,
> That in summer may burn, and in winter may splash;
> I love dust and dirt; and 'tis always my pleasure,
> To take with me much of the soil that I measure.
>
> "But Matthew thought better: for Matthew thought right,
> And hired a chariot so trim and so tight,
> That extremes both of winter and summer might pass:
> For one window was canvas, the other was glass."

Prior evidently liked the chaises of Holland.

> "While with labour assiduous due pleasure I mix,
> And in one day atone for the business of six,
> In a little Dutch chaise on a Saturday night,
> On my left hand my Horace, a nymph on my right:
> No Memoirs to compose, and no Post-boy to move,
> That on Sunday may hinder the softness of love;
> For her, neither visits, nor parties at tea,
> Nor the long-winded cant of a dull Refugee:
> This night and the next shall be hers, shall be mine,
> To good or ill-fortune the third we resign:
> Thus scorning the world and superior to fate,
> I drive on my car in processional state."

Another advertisement tells of a gentleman who brought a one-horse calash to an Inn near Hyde Park Corner, took away the horse ten days later, but left his carriage "as a pawn for what was due for the same." In a while the inn-keeper was advertising the fact that unless the owner claimed it within ten days he should sell the carriage for what it would fetch. A more curious advertisement belonging to this period may be quoted in full:—

> "Lost the 26th of February, about 9 a Clock at Night, between the Angel and Crown Tavern in Threadneedle Street, and the end of Bucklers Berry, the side door of a Chariot, Painted Coffee Colour, with a Round Cypher in the Pannel, Lin'd with White Cloath embos'd with Red, having a Glass in one Frame, and White Canvas in another, with Red Strings to both Frames. Whoever hath taken it up are desir'd to bring it to Mr. Jacob's a Coachmaker at the corner of St. Mary Ax near London Wall, where they shall receive 30s. Reward if all be brought with it; or if offer'd to be Pawn'd or Sold, desire it may be stop'd and notice given, or if already Pawn'd or Sold, their money again."

At this time, if not before, it became customary for wealthy people to possess coaches used only when they were in mourning. So we have:—

> "At Mr. Harrison's, Coach Maker, in the Broadway, Westmin-

ster, is a Mourning Coach and Harness, never used, with a whole Fore Glass, and Two Glasses and all other Materials (the Person being deceased); also a Mourning Chariot, being little used, with all Materials likewise, and a Leather Body Coach, being very fashionable with a Coafoay Lining and 4 Glasses, and several sorts of Shazesses, at very reasonable rates."

What these reasonable rates were does not appear, but we learn from an agreement made in 1718 between one Hodges, a job-master, and a private gentleman, the cost of hiring a complete equipage. Hodges was to maintain "a coach, chariot, and harness neat and clean, and in all manner of repair at his own charge, not including the wheels, for a consideration of five shillings and sixpence a day—this to include a pair of well-matched horses and a good, sober, honest, creditable coachman." If extra horses were required for country work, they were to be had for half a crown the pair per day. And if the coachman should break the glass when the coach was empty, Hodges and not the private gentleman should be responsible for the damage.

From another advertisement of about the same time comes the information that the hammercloth of carriages was constantly being stolen. Ashton[37] gives three such advertisements.

"Lost off a Gentleman's Coach Box a Crimson Coffoy Hammer Cloth, with 2 yellow Laces about it."

"Lost off a Gentleman's Coach Box, a Blue Hammer Cloth, trimm'd with a Gold colour'd Lace that is almost turn'd yellow."

"Lost a Red Shag Hammock Cloth, with white Silk Lace round it, embroider'd with white and blue, and 3 Bulls Heads and a Squirrel for the Coat of Arms."

The etymology of this hammercloth, which was simply a covering over the coach-box, seems to have puzzled people considerably. Most coachbuilders consider that the box beneath the seat used to contain a hammer and other tools necessary in case of a breakdown, whence the name. The anonymous author of the coach-building articles in the *Carriage Builders' and Harness-Makers' Art Journal* scouts this idea, and suggests that it is merely a corruption of hamper-cloth—the box or chest having originally contained a hamper of provisions. The last advertisement quoted above gives hammock-cloth, which vaguely suggests suspension of a kind. It is perhaps not a very important question.

Advertisements also mention a "*Curtin Coach for 6 People,*" and "a *Chasse marée Coach,*"[38] which was some form of covered waggon; but, unfortunately, I have not been able to discover any information about them.

---

[37] *Social Life in the Reign of Queen Anne.* John Ashton. London, 1883.
[38] Originally, I understand, a fish-cart or lugger.

Queen Anne's Procession to the Cathedral of S. Paul

(From Pennant's "London")

The State Coaches of this time were as handsome as ever. George I, Mrs. Delaney has recorded, rode in a coach that was "covered with purple cloth; the eight horses the beautifullest creatures of their kind were cream colour"—the custom of using cream-coloured horses still obtains in the State Coach of Great Britain—"the trapping purple silk, and their manes and tails tied with purple riband." Luttrell in his Diary for May 20th, 1707, says of a foreign coach:—

> "Yesterday the Venetian Ambassadors made their public entry thro' this citty to Somerset House in great state and splendour, their Coach of State embroidered with gold, and the richest that ever was seen in England: they had two with 8 horses, and eight with 6 horses, trimm'd very fine with ribbons, 48 footmen in blew velvet cover'd with gold lace, 24 gentlemen and pages on horseback, with feathers in their hats."

The Venetians apparently prided themselves on a magnificent display, and four years later Swift, in one of his letters to Stella, was commenting upon their ambassador's coach again—"the most monstrous, huge, fine, gilt thing that ever I saw," he says of it. Every possible luxury was commandeered for these State vehicles. One of the Emperors built a coach "studded with gold" for his bride. Another's consort rode in a carriage "covered with perfumed leather." The wedding carriage of the first wife of the Emperor Leopold had cost 38,000 florins. But the Austrian State Imperial Coach, built in 1696, was perhaps the most gorgeous of all. Immense sums too were being spent on coaches by private individuals. Swift writes on February 6th, 1712: "Nothing has made so great a noise as one Kelson's chariot, that cost nine hundred and thirty Pounds, the finest was ever seen. The rabble huzzaed him as much as they did Prince Eugene." Fashion decreed six horses. "I must have Six Horses in my Coach," says Mrs. Plotwell in the *Beau's Duel*, "four are fit for those that have a Charge of Children, you and I shall never have any"; and in another of Mrs. Centlivre's comedies, Lucinda says to Sir Toby Doubtful: "You'll at least keep Six Horses, Sir Toby, for I wou'd not make a Tour in High Park with less for the World: for me thinks a pair looks like a Hackney." Abroad even more display was made. "Two coaches," wrote Lady Mary from Naples in 1740, "two running footmen, four other footmen, a gentleman usher, and two pages, are as necessary here as the attendance of a single servant is at London."

Nor was carriage-driving confined to the gentry. Every retired tradesman appeared abroad in his coach and aped the noble, a matter which disturbed Sir Richard Steele, who in one of the *Tatlers* drew attention to the truly lamentable fact that you could not possibly estimate the social position of the occupant of a coach by the appearance of his equipage.

> "For the better understanding of things and persons," he writes, "in this general confusion, I have given directions to all the coachmakers and coachpainters in town, to bring me in lists of their several customers; and doubt not, but with comparing the orders of each man, in the placing of his arms on the door of his chariot, as well as the words, devices and ciphers to be fixed upon them, to make a collection which shall let us into the nature, if

not the history, of mankind, more usefully than the curiosities of any medallist in Europe. It is high time," he continues, "that I call in such coaches as are in their embellishment improper for the character of their owners. But if I find I am not obeyed herein, and think I cannot pull down those equipages already erected, I shall take upon me to prevent the growth of this evil for the future, by inquiring into the pretensions of the persons, who shall hereafter attempt to make public entries with ornaments and decorations of their own appointment. If a man, who believed he had the handsomest leg in this kingdom, should take a fancy to adorn so deserving a limb with a blue garter, he would be justly punished for offending against the Most Noble Order; and, I think, the general prostitution of equipage and retinue is as destructive to all distinction, as the impertinences of one man, if permitted, would certainly be to that illustrious fraternity."

The temptation for display must have been great. Nothing attracted the public attention like a fine coach. In the north of Scotland, indeed, any carriage caused the profoundest astonishment.

"I was entertained," says a contemporary writer, "with the Surprise and Amusement of the Common People when in the year 1725 a Chariot with six monstrous great Horses arrived here by way of the Sea Coast. An Elephant publicly exhibited in the Streets of London could not have excited greater admiration. One asked what the Chariot was; another, who had seen the gentlemen alight, told the first with a Sneer at his Ignorance, it was a great cart to carry people in, and such like."

And even in Johnson's day, when there were few coaches to be found in this part of the country, though a lighter vehicle called in old account books a *cheas* was sometimes used, public astonishment was great. Yet it was in the north of Scotland that military roads were constructed in 1726 and 1737—not particularly good roads, but very necessary—and the first of their kind.

Swift in *Apollo, or a Problem Solved*, satirised the prevailing luxury. Compared with Apollo, he says:—

"No heir upon his first appearance,
With twenty thousand pounds a year rents,
E'er drove, before he sold his land,
So fine a coach along the Strand:
The spokes, we are by Ovid told,
Were silver, and the axle gold:
I own, 'twas but a coach-and-four,
For Jupiter allows no more."

But whether Jupiter allowed it or not, your fashionable dame had six horses put into her coach, and the more grooms in attendance upon her, the better for her reputation as a Person of Quality. There is a good story, by the way, of Swift and a hackney coach. It is told by Leigh Hunt in his essay on *Coaches*.

"He was going," says Hunt, "one dark evening, to dine with some great man, and was accompanied by some other clergymen, to whom he gave their clue. They were all in their canonicals. When they arrive at the house, the coachman opens the door, and lets down the steps. Down steps the Dean, very reverend in his black robes; after him comes another personage, equally black and dignified; then another; then a fourth. The coachman, who recollects taking up no greater number, is about to put up the steps, when another clergyman descends. After giving way to this other, he proceeds with great confidence to toss them up, when lo! another comes. Well, there cannot, he thinks, be more than six. He is mistaken. Down comes a seventh, then an eighth; then a ninth; all with decent intervals; the coach in the meantime rocking as if it were giving birth to so many daemons. The coachman can conclude no less. He cries out 'The devil! the devil!' and is preparing to run away, when they all burst into laughter. They had gone round as they descended, and got in at the other door."

It may be that the private coaches and chariots were rather more comfortable than the hackneys, but nothing, it seems, could equal the tortures which were inflicted upon the unfortunate passengers who were forced to ride in the public carriages.

"When our *Stratford* Tub," writes Ned Ward, "by the Assistance of its Carrionly Tits of different colours, had outrun the Smoothness of the Road, and enter'd upon *London*-Stones, with as frightful a Rumbling as an empty *Hay-Cart*, our Leathern-Conveniency[39] being bound in the Braces to its good Behaviour, had no more Sway than a *Funeral Hearse*, or a *Country-Waggon*, that we were jumbled about like so many *Pease* in a Childs-Rattle, running, at every Kennel-Jolt, a great hazard of a Dislocation: This we endured till we were brought within White-Chappel Bars, where we Lighted from our Stubborn Caravan, with our Elbows and Shoulders as Black and Blew as a *Rural Joan*, that had been under the Pinches of an Angry *Fairy*. Our weary Limbs being rather more Tir'd than Refresh'd, by the Thumps and Tosses of our ill-contriv'd Engine, as unfit to move upon a Rugged Pavement as a Gouty Sinner is to valt o'er *London Bridge*, with his Boots on. For my part, said I, if this be the Pleasure of Riding in a Coach thro' *London-Streets*, may those that like it enjoy it, for it has loosen'd my Joynts in so short a Passage, that I shall scarce recover my former Strength this Fortnight; and, indeed, of the two, I would rather chuse to cry *Mouse-Traps* for a Livelihood, than be oblig'd

---

[39] This well-known expression for a carriage is generally thought to have been used first by an American quaker later in the century. Ned Ward, however, would seem to have been its real inventor.

every day to be drag'd about Town under such uneasiness; and if the Qualities Coaches are as troublesome as this, I would not be bound to do their Pennance for their Estates. You must consider, says my Friend, you have not the right Knack of Humouring the Coaches Motion; for there is as much Art in Sitting in a coach finely, as there is in riding the Great Horse; and many a younger Brother has got a good Fortune by his Genteel Stepping in and out, when he pays a Visit to her Ladyship."

In Fleet Street, it seems, things were very bad. "The Ratling of Coaches," says Ward, "loud as the *Cataracts* of *Nile* Rob'd me of my Hearing, and put my Head into as much disorder as the untunable Hollows of a Rural Mob at a Country *Bull-Baiting*." More trouble followed later in the day.

"Now, says my Friend, I believe we are not tired with the Labours of the Day; let us therefore Dedicate the latter part purely to our Pleasure, take a Coach and go see *May-Fair*. Would you have me, said I, undergo the Punishment of a Coach again, when you know I was made so great a sufferer by the last, that it made my Bones rattle in my Skin, and has brought as many Pains about me, as if troubled with Rheumatism. That was a Country Coach, says he, and only fit for the Road; but *London* Coaches are hung more loose to prevent your being Jolted by the Roughness of the Pavement. This Argument of my Friends prevail'd upon me, to venture my Carcase a second Time to be Rock'd in a *Hackney* Cradle. So we took Leave of the *Temple*, turn'd up without *Temple-Bar*, and there took Coach for the General Rendezvous aforementioned.

"By the help of a great many Slashes and Hey-ups, and after as many Jolts and Jumbles, we were dragg'd to the *Fair*, where our Charioteer had difficulty with his fare—the gay ladies refusing to pay, but one eventually pledging her scarf and taking his number."

It is to be remembered that at this time, as in the last century, the hackney coaches were used much in the manner of the modern omnibus. You did not necessarily have one to yourself. The same held good with regard to the post-chaises. Advertisements were constantly appearing for a "partner."

The uneasy motion which so disturbed Ned Ward was a matter which was receiving the attention of carriage-builders, but little enough was done. Yet in England, France and Spain, quite a number of strange machines (including one which was supposed to go without horses) were invented, and had their day, and disappeared into the lumber-room of time. Two in particular, though in the main unsuccessful, deserve mention.

One, properly belonging to the seventeenth century, concerned a new steel spring, patented in 1691 by a Mr. John Green. It was thus advertised:—

"All the nobility and gentry may have the carriages of their coaches made new or the old ones altered, after this invention, at reasonable rates; and hackney and stage coachmen may have licences from the Patentees, Mr. *John Green* and Mr. *William Dockwra*, his partner, at the rate of 12*d.* per week, to drive the roads and streets, some of which having this week began, and may be known from the common coaches by the words patent Coach being over both doors in carved letters. These coaches are so hung as to render them easier for the passenger and less labour for the horses, the gentleman's coaches turning in narrow streets and lanes in as little or less room than any French carriage with crane neck, and not one third of the charge. The manner of the coachman's sitting is more convenient, and the motion like that of a sedan, being free from the tossing and jolting to which other coaches are liable over rough and broken roads, pavements or kennels. These great Conveniences (besides others) are invitation sufficient for all persons that love their own ease and would save their horses draught, to use these sort of carriages and no other, since these carriages need no alteration."

Here, in addition to the spring, there was some kind of turning head—a question which occupied the attention of designers throughout the next century, but nothing more of Mr. John Green or of his partner was heard of, and his patent coaches found few if any purchasers.

The other contrivance was a primitive form of gear invented by one James Rowe. In 1727 this Rowe wrote a book—not, however, published until 1734—called *All Sorts of Wheel Carriage, Improved.* This was a small tract "wherein is plainly made to appear, that a much less than the usual Draught of Horses, etc., will be required, in Waggons, Carts, Coaches, and all other Wheel Vehicles" by the application of small "friction wheels and pulleys." Rowe obtained a patent for his gear and apparently applied his small wheels to the axle just within the ordinary wheels, but his own coach was probably the only one ever to be so fitted. It was felt no doubt that the whole question was one of roads rather than of carriages. Improve your roads, and the discomforts of travelling would disappear.

The British stage-coaches of this time were, according to Sir Walter Scott,

"constructed principally of a dull black leather, thickly studded, by way of ornament, with black-headed nails tracing out the panels; in the upper tier of which were four oval windows, with heavy red wooden frames, and green stuff or leathern curtains. Upon the doors, also, there appeared but little of that gay blazonry which shines upon the numerous quadrigae of the present time; but there were displayed in large characters the names of the places whence the coach started, and whither it went, stated in quaint and ancient language. The vehicles themselves varied in shape. Sometimes they were like a distiller's vat; sometimes flattened, and hung equally balanced between the immense front and back springs; in other instances they resembled a violincello case, which was past all comparison the most fashionable form;

and they hung in a more genteel posture, namely, inclining on to the back springs, and giving to those who sat within the appearance of a stiff Guy Faux, uneasily seated. The roofs of the coaches, in most cases, rose into a swelling curve, which was sometimes surrounded by a high iron guard.... The coachman, and the guard, who always held his carabine ready bent, or, as we now say, cocked upon his knee, then sat together; not as at present, upon a close, compact varnished seat, but over a very long and narrow boot, which passed under a large spreading hammer cloth, hanging down on all sides, and finished with a flowing and most luxurious fringe. Behind the coach was the immense basket stretching far and wide beyond the body, to which it was attached by long iron bars or supports passing beneath it; though even these seemed scarcely equal to the enormous weight with which they were frequently loaded. They were, however, never very great favourites, although their difference of price caused them frequently to be well filled, for, as an ancient Teague observed, 'they got in so long after the coach, that they ought to set out a day sooner, to be there at the same time. Arrah!' continued he, 'can't they give it the two hind wheels, and let it go first?' The wheels of these old carriages were large, massive, ill-formed, and usually of a red colour; and the three horses that were affixed to the whole machine—the foremost of which was helped onward by carrying a huge long-legged elf of a postillion, dressed in a cocked hat, with a large green and gold riding coat—were all so far parted from it by the great length of their traces, that it was with no little difficulty that the poor animals dragged their unwieldy burthen along the road. It groaned, and creaked, and lumbered, at every fresh tug which they gave it, as a ship, rocking or beating up, through a heavy sea, strains all her timbers with a low-moaning sound, as she drives over the contending waves."

No wonder, said Scott, that at this time people invariably made their wills before setting out on a journey of any length. The dangers were manifold and very real.

In France the stage-coaches, or *diligences*, were very similar "with large bodies, having three small windows on each side and hung by leather braces on long perch carriages, with high hind wheels and low front wheels, without any driving box and fitted with large baskets, back and front for passengers or luggage; they were drawn by five horses and driven by a postillion on the off wheeler instead of the near wheeler as in England." One, at any rate, of these diligences had springs of a kind. Another public coach in France at this time was the *gondola*, holding ten or twelve passengers inside, these sitting sideways with one at each end, a second attempt at a kind of omnibus. Still another public vehicle popular about this time in Paris was the *coucou*. Of this weird machine Ramée says:—

"Figure a box, yellow, green, brown, red, or sky blue, open in front, having two foul benches which had formerly been stuffed, on which were placed six unfortunate voyagers. In the sides it had, right and left, one or two square openings, to give air during the day or in summer. While the interior was sufficiently open to the world, there was built an apron in front, framed in woodwork and covered with sheet iron. Upon this apron was thrown a third bench, on which were seated the driver of the *coucou* and two passengers who were termed *lapins* (rabbits)."

The *coucou* was regularly to be seen lumbering painfully along with its ten or a dozen passengers, its snail's pace giving it the ironical name of *vigoureux*. The poorer people almost exclusively used the *coucou*, although a smart woman with her pet dog, or a gentleman who had been unable to find a place in the more aristocratic *gondola*, were occasionally to be seen in its interior sandwiched in between two peasants.

In Spain the *coucou* found an equivalent in the *galera*, which was provided with the ubiquitous basket—a low waggon it was, with its sides formed of a number of wooden spokes at a considerable distance from each other, and having no bottom save a strip of spartum on which the trunks and packages were heaped. In Spain there were several types of cart, two-or four-wheeled, which likewise plied for passenger hire. One of these, called a *correo real*, seems to have travelled at rather a greater pace, though with even less comfort to the unfortunate passengers than the others. A century later this *correo real* was described by Théophile Gautier, who speaks of it as "an antediluvian vehicle, of which the model could only be found in the fossil remains of Spain, immense bell-shaped wheels, with very thin spokes, considerably behind the frame, which had been painted red somewhere about the time of Isabella the Catholic; an extravagant body full of all sorts of crooked windows, and lined in the inside with small satin cushions, which may at some period have been rose-coloured, and the whole decorated with a kind of silk that was once probably of various colours."

In 1743 the system of travelling post, which so long before as 1664 had been common in France, was introduced into England by one John Trull, an artillery officer, who obtained a patent for letting carriages for hire across country. These were the *post-chaises*, of which the first were two-wheeled with the door in front— in this respect being similar to the French *chaises de poste*, from which the idea was taken. Trull's scheme, however, though successful in itself does not seem to have brought money to its inventor, who thirty years later died in the King's Bench. The door of these first post-chaises "was hinged at the bottom and fell forward on to a small dasher like a gentleman's cabriolet," and there was a window on either side. "It was hung upon two very lofty wheels," says Thrupp, "and long shafts for one horse, and the body was rather in front of the wheels, so that the weight on the horse's back must have been considerable. It was suspended at first upon leather braces only, but later upon two upright or whip springs behind, and two elbow springs in front from the body to the cross-bar, which joined the shafts and carried the step." Soon, however, these post-chaises were built with four wheels, and resembled the ordinary private chariots of the day, though without their lavish ornamentation. In less than ten years, however, a larger body was given to them, so that they came to resemble the coach rather than the smaller and slimmer chariots, while the coachman's box was made very much higher.

The post-chaise became extraordinarily popular. The literature of the mid-eighteenth century is full of references to it. All kinds of adventures happened to people in post-chaises. They were seen in every part of the country, they could be hired here, there, and everywhere. Dr. Johnson was only one amongst thousands who loved them. "If I had no duties," he records, "and no reference to futurity, I would spend my life in driving briskly in a post-chaise with a pretty woman." "I have tried almost every mode of travelling since I saw you," wrote Wilkes to his daughter, "in a coach, chaise, waggon, boat, *treckscuyt*, *traineau*, sledge, etc. I know none so agreeable as my English post-chaise."

One thinks naturally of Laurence Sterne. Both in *Tristram Shandy* and in the *Sentimental Journey* he has much to say of the post-chaises. "Something is always wrong," he is grumbling somewhere, "in a French post-chaise, upon first setting out.... A French postillion has always to alight before he has got three hundred yards out of town." And then, of course, there is that never-to-be-forgotten *désobligeante* which he purchased from M. Dessein at Calais.[40]

> "There being no travelling in France and Italy," he recounts, "without a chaise—and nature generally prompting us to the thing we are fittest for, I walk'd out into the coach-yard to buy or hire something of that kind to my purpose: an old *Désobligeante*, in the furthest corner of the court, hit my fancy at first sight, so I instantly got into it."

And there it was in that queer little carriage which would hold but one person, that Sterne wrote his famous Preface about Travellers, "though it would have been better," he observed, when interrupted, "in a *Vis-à-Vis*." The particular *désobligeante* seems to have proved satisfactory, but for the species Sterne could not find much praise.

> "In Monsieur Dessein's coach-yard," he says, "I saw another old tatter'd *désobligeante*; and notwithstanding it was the exact picture of that which had hit my fancy so much in the coach-yard but an hour before, the very sight of it stirr'd up a disagreeable sensation within me now; and I thought 'twas a churlish beast into whose heart the idea could first enter, to construct such a machine; nor had I much more charity for the man who could think of using it."

---

[40] At this time M. Dessein used to advertise in the London papers. In *The Gazetteer and New Daily Advertiser* for July 21, 1767, is the following: "To be sold, at Calais, a Travelling *Vis-à-Vis*, built at Paris about a year and a half ago; very fit also to use in the towns on the Continent upon occasion; being varnished in the newest taste, and covered with an oiled case to preserve it from the weather in travelling, and requires nothing but a new set of wheels to be in perfect repair to make the tour of Europe. Enquire of Mr. Dessein, at the Hôtel D'Angleterre at Calais, with whom the lowest price is left."

It was certainly not a very sociable carriage, but then neither was the sedan: both were very useful.

I may conclude this chapter by drawing attention to the tax upon coaches which was levied at the beginning of 1747. From the fuss that was made when such a bill was first introduced—it was temporarily abandoned—you might imagine that one of the most treasured articles of the Constitution was about to be swept away.

"It is impossible to express," wrote a country clergyman to his bishop in a letter which deserves quotation as affording an insight into the lesser equipages used in the country at this time, "the various impressions your lordship's letter, relating to the tax upon coaches, made here; as people imagined it a jest, or serious: As most inclined to the former, it would be too tedious to trouble you with the witticisms and conundrums it occasioned. B. said the Church was in danger; C. observed it would be like the gospel-feast inverted, that the maimed and lame being the only guests admitted there, would be the only ones excluded here.... As we have now no reason to doubt such a tax being really intended, give me leave to represent to you our thoughts of it here. My living, your Lordship knows, is under £70 *per Ann.*, yet out of this, some years since, I made a shift to lay out six pounds on an old chariot, which, with the help of my ploughman and a pair of cart-horses, has drawn my wife, etc., half a mile to church, who, for the future, must go in a cart, or stay at home. Repairs, etc., have cost me, *communibus annis*, for the eleven years I have had it, about 7s. so the interest of my money, at 5 *per cent*, on the £6 and 7s. in repairs, is 13s. *per Ann.*, which with tax on this my pompous luxury, will be increased to £4 13s. *per Ann.*, almost the prime cost of setting up my Equipage. I am afraid this is not my case singly, but will be found pretty nearly so, of most of the small clergy in England. Among the laity we have several gentlemen farmers, who manage, in some degree, with the same frugality, and who, for the same reasons, are prepared to part with, or continue them according to the fate of this bill; insomuch, that I can compute that in sixteen parishes I have in my eye three times that number of coaches will be disposed of, for we look on the same sum, which is but a trifling duty on grand equipages, to amount to a prohibition on ours, which resembles them no more than a ragged coat does an embroidered suit. I shall not dwell on the quantity of glass (not to mention leather, etc.), this will bring to market, nor the future consumption of these commodities it will prevent.... To me I own it looks a little like the son eating the father.... How many single gentlemen," he goes on to ask, after pointing out that it is the poorer married men who will suffer most, "from 2, 3, to 800*l.* a year, and more, have no coaches, yet keep a stable of hunters (the worst of which would purchase my equipage) and a pack of hounds, whom this duty will not affect?"

But the bill was passed, and so we must suppose that our clergyman and his farmer friends were forced to walk to church.

Some verses printed in the *Gentleman's Magazine* at this time may also be quoted as reflecting the general opinion about the bill.

"Before Bohemian *Anne* was Queen,
Astride their steeds were ladies seen;
And good Queen *Bess* to *Paul's*, I wot,
Full oft astride has jogg'd on trot:
Beaus then could foot it thro' all weather,
And nothing fear'd but wear of leather.
But now (so luxury decrees)
The polish'd age rolls on at ease:
Coach, chariot, chaise, berlin, landau[41]
(Machines the ancients never saw)
Indulge our gentle sons of war,
Who ne'er will mount triumphant car.
The carriage marks the peer's degree,
And almost tells the doctor's fee;
Bears ev'ry thriving child of art;
Ev'n thieves to *Tyburn* claim their cart.
O cruel law! replete with pain,
That makes us use our feet again;
Or, half our pair oblig'd to lack,
Bids us bestride the other's back.
A shilling stage would suit with many,
Who cannot reach an eighteen penny.
Rock must enhance the price of pills,
Or drive again—one pair of wheels.
The graduate too will be to seek,
Who mounts his chariot twice a week:
For if the hackneymen should grumble,
I fear our Phaeton[42] must tumble.
O cruel law! to raise the fare
Of Christmas turkey, chine, and hare;
The vails or wages to retrench
Of country serving-man or wench,
Who twice a year ride up and down,
Betwixt their native place and town.
O cruel tax! who must not say,
Which only those who will—need pay?"

From this bill, those who used the one-horse chaises certainly suffered. *Rusti-*

---

[41] See next chapter.
[42] See next chapter.

cus thereupon offered the following advice to his fellow-sufferers at the time of the next General Election:—

> "Ye who late loll'd in easy *chaise and one,*
> And now must walk, or ride *Old Grey* or *Dun,*
> Enquire when wheels were tax'd (to mend your fate)
> What patriots, *spokesmen* were in the debate.
> And get this act, a promise to revoke,
> Or *put* into each *spokesman's* wheel a *spoke*."

CHAPTER THE SEVENTH

THE WAR OF THE WHEELS: WITH SOME CURIOSITIES, REGAL AND OTHERWISE

"The morning came, the chaise was brought,
But yet was not allowed
To drive up to the door, lest all
Should say that she was proud.

"So three doors off the chaise was stayed,
When they did all get in,
Six precious souls, and all agog
To dash through thick and thin."

John Gilpin.

"IN my journey to *London*," wrote an indignant correspondent to the *Gentleman's Magazine*, early in 1747, "I travell'd from *Harborough* to *Northampton*, and well was it that I was in a light Berlin, and six good horses, or I might have overlaid in that turnpike road. But for fear of life and limb, I walk'd several miles on foot, met 20 waggons tearing their goods to pieces, and the drivers cursing and swearing for being robb'd on the highway by a turnpike, screen'd under an Act of Parliament." These turnpikes, or toll-gates, had been but lately established in England for the preservation of the roads. That they did very much immediate good, however, may be doubted. A few years afterwards an English traveller was grumbling at the superiority of the French roads over our own. "Nothing piques me more," he wrote in an amusingly satirical passage, "than that a trumpery despotic Government, like France, should have enchanting roads from the capital to each remote part." He seems to have taken trouble to find out the cause for so lamentable a difference, and for this purpose consulted "the most solemn looking waggoner on the road."

"This prov'd to be *Jack Whipcord* of *Blandford*. Jack's answer was 'That roads had but one object, namely, waggon-driving. That he requir'd but five feet wedth in a line [which he resolved never to quit], and all the rest might go to the d— — l. That the gentry ought to stay at home and be d— — and not run gossiping up and down the country. But, added *Jack*, we will soon cure them, for my brethren since the late act have made a vow to run our wheels in the coach quarter. We tack on a sixth or seventh horse at pleasure. What a plague would they send us to the galleys for this, as papishes do in beyond-sea countries.'"

The Act to which Jack referred had been passed in 1745. It followed upon the fact that while coaches, generally speaking, were in process of becoming lighter, carts and waggons were becoming much heavier. And so it had been proposed that no waggon should be drawn by more than four horses, no matter whether these were "in length, pairs or sideways," and no cart should have more than three. Every horse above these numbers could be forfeited together "with all geers, bridles, halters, harness and accoutrements." There were to be collectors of tolls, and gentlemen's private carriages and purely agricultural waggons were to be exempt. Also certain roads, presumably those but lately laid down according to the best ideas of the time, were to be treated as outside the scope of the Act. And if the wheels of these heavy waggons and carts possessed "wheels bound with streaks or tire of the breadth of eight inches at least when worn and not set on with rose-headed nails," they might likewise be exempt.

This Bill gave rise to a curious wordy warfare, which was carried on for some years, and may be said to have interested people in the general questions of wheeled traffic right on until the time when McAdam's schemes altogether altered general opinion. This war, of course, hardly touched private carriages, but was waged in so many quarters and with such various weapons that it deserves some mention in any account of carriages.

It was immediately "objected by multitudes" that the Bill of 1745 would "greatly enhance the price of carriage of goods," but its apologists argued that even if it did, better-designed carriages and carts would be built, so that the roads would improve, and the price of cartage ultimately go down. "It is urged," they said, "that light carts or waggons may be used, and the horses draw double, as in the rabbet waggons of Norfolk, which improves the road and contributes to expedition."

At an early stage in this war two factions arose. On the one hand you had coachbuilders and others filling the newspapers and publishing tracts, some very serious, some extraordinarily mathematical, others merely facetious, to prove that the roads could be preserved only by using very broad wheels—some, indeed, advocated rollers, which, as we shall see, were actually tried—and on the other hand you had people filling more columns, and very dull columns some of them were, to show that a low broad wheel was the one thing which no really satisfactory vehicle could possibly possess. These were the apologists for the lighter waggons with large but slender wheels. Decrease your weight, said they, and never mind about the wheels; it is the great weight that ruins the roads. How can you decrease the weight, asked the broad-wheel faction, without increasing the cost of carriage? Increase the cost of carriage for a while, was the reply, and see what happens to the roads.

For a time, however, the broad-wheel faction held the advantage, and when further legislation was made in 1754, it was entirely in their favour.

> "It is enacted that after next Michaelmas, no wheel carriage of burthen (except it be drawn by oxen only, or if by horses with less than five, if a four-wheeled carriage, with less than four) shall travel any turnpike road, unless the fellies of the wheels shall be nine inches from side to side under a penalty of £5, or the forfeiture of one of the horses, with all his accoutrements, to the sole use of the person who shall seize them."

So soon as such proposals had become law, it was asked with some pertinence: where were these huge wheels to come from? What of the heavy expenses that would fall on the farmers? The parrot cry, "Your wheels will cost you more," was hinted at, if not expressed in so modern a way. Arguments were put forward to show that the correct height for wheels was anything between two and eight feet, and the correct breadth from three to eighteen inches. And the disputes became tinged with personalities. But the net result seems to have been that most people fought shy of the very wide wheels, and were content to use less horses.

The war dragged on, and particular inventions to cope with the difficulty began to appear. A new tire was widely advertised. An enthusiastic inventor occupied two or three pages of the *Gentleman's Magazine* with details of his particular waggon, which had the front and back wheels of very different sizes, but what exactly its advantages might be were not very clear to any one but himself. Then on the 14th of April, 1764, one Daniel Bourn of Leominster produced a waggon on small rollers. Though it was unsuccessful, it led the way to further experiments, and as will be seen from the contemporary account immediately below, contained at any rate one novel feature which was subsequently widely adopted not only in waggons and carts, but also in four-wheeled carriages of every description.

> "Mr. Bourn's new machine for travelling the roads was tried against a common broad-wheeled waggon, but did not answer, the common waggon going as well with four horses, as the new one with eight. The weight carried was five ton besides the carriage. The wheels of this waggon are 14 inches; the fore wheels go within the hind wheels, and are so shallow as to turn under the bed of the waggon. The *Leominster* stage waggon has these wheels."

The experiment took place "abreast between the new road just by Pancras to within a small distance of Bog-house Bar." Apparently the only advantage which the new waggon possessed was its ability to turn in a narrow road, but although Mr. Bourn not only continued to build such waggons, but also answered his opponents in two tracts, we hear little more of him. Such "rolling-carts," however, were also made by one James Sharpe, of Leadenhall Street. Sharpe was a pushful man. He believed in his system, and apparently made those in authority see its advantages. His rollers, you learn, were cylinders of cast iron, two feet in diameter and sixteen inches broad. An iron spindle was inserted through the centre of each. Several of Sharpe's waggons were on the roads, but although every facility was given them, they never really took the popular fancy. And indeed, they must have been uncouth monsters rattling along the roads something after the fashion of the steam-roller of ten years ago. Just about this time, too, the light-cart faction showed that it was not in the least moribund. It indited learned and highly technical articles which the newspapers found space to print with some regularity. A typical reply to such articles was inserted in the *Public Advertiser* early in 1767:—

"There are people, I may say," runs this most impolite retort, "a depraved Number, who write long letters upon this Subject in an ignorant Manner. Their Errors confirm Mankind of a sensible Turn what Measures ought to be taken for the Benefit of the trading Part of this Nation. The illiterate Scribblers he [the last correspondent] means to lash are those that insist upon the Necessity of Horses going at length instead of being placed abreast. The Power that draft Horses have in being placed abreast is so well known, that 'tis amazing any body is absurd enough to advance a Doctrine to the contrary. Then again, these deluded Idiots propagate, that the Loads drawn by eight Horses, having the Wheels placed nine Inches within nine, are destructive to the Roads, and that the Weight had better be divided into several narrow wheeled Carriages. Being thus destitute of Judgment upon the Subject, they do not reflect that the more Horses and Carriages the more the Expence increases, consequently that the internal Trade of this Kingdom would advance in this Article 100 per Cent. One Waggon with eight Horses in Pairs, drawing eight Ton upon the new Plan, don't do near the Mischief that the same Weight would in two Waggons with narrow Wheels. Besides, four Horses at length cannot draw four Ton Weight. A late trifling Writer upon the Subject says, the Appearance of a broad Wheel Waggon was terrific. I think he may be pronounced a Cockney without Ceremony—a Cit that carries his Wife and Children four Miles out of Town in a Tim-Whisky, and, being most likely an aukward Driver, suffers the Squalls of his Horn-making Spouse to alarm his Dove-like Pusillanimity."

Such a man, the article goes on to say, would surely be frightened if he saw a three-master sailing the seas, and he and his kind had better keep quiet upon a subject of which they appeared so entirely and pitiably ignorant.

The contest began to embrace wider issues than the mere wheels of waggons. It took in the whole question of wheeled carriages. It even went so far as to include a denunciation of the general policy of the Government, whose legislation, or lack of it, on this vexed question was, so the light-cart faction maintained, leading directly to an increase in the price of provisions. Nothing, apparently, was right. If waggons were constructed on principles which were as bad as they could be, so were the Stage-coaches, which also were using the public roads, though some of the controversialists seemed to forget the fact.

"We are desired," runs a paragraph in the newspapers of this time, "to inform the Masters of Stage-Coaches, Machines, &c., that their present Method of hanging their Carriages high with a low Fore-Wheel, and the body of the Coach hung forwards with the Stems of the Box leaning likewise forwards, is all upon a ridiculous wrong Principle,—the Effect of the Stupidity of Coachmen and Wheelwrights; that if they pursue the following Regulations, they will find the same Advantage that the Nobility and Gentry have already done by adopting this Plan: Let the fore Wheels be three Feet, six, eight, or ten Inches high, the Stems of the Box upright, and admit as little Weight forward as possible upon the low Wheel; the Body of the Coach to hang low for the Convenience of Passengers, as no Benefit arises from its Hanging high to the Horses, their Advantage laying intirely upon the Height of the Fore-Wheels."

This in its turn was argued. Then came a proposal to tax private carriages according to the number of horses used, and see whether such revenue would not counterbalance in some way the increase in the prices of provisions, which, of course, was following on this eternal wrangle of the waggons. Also there was more legislation. Some of the new regulations read curiously. "No tree or bush is to be allowed to grow or stand within fifteen feet of the center of the highway, on forfeiture of 10s. by the owner." Cartways were to be at least twenty feet wide, and horse causeways three feet wide. No waggon with more than four horses might have wheels less than nine inches in width, and some one on horseback or on foot had to go in front of it. More criticism filled the newspapers, and more inventions appeared.

Meetings were held. One advertisement which appeared in 1767 has an agreeable air of mystery about it.

"All persons working Shod-wheel'd Carts, Waggons, Drays, &c. of all Breadths, are desired to meet at the Sun Tavern in St. Paul's Churchyard, on Friday next, at four o'clock in the Afternoon. Enquire for No. 1."

And more pamphlets appeared, but the roads failed to improve.

Then in 1770 another Act was passed giving privileges to the roller-carts which were denied to the ordinary waggons. "All carriages," it ordered, "moving upon rollers the breadth of fifteen inches, are allowed to be drawn with any number of horses, or other cattle." And, as a further inducement, such carts were to be toll-free for a year. Mr. Sharpe, of Leadenhall Street, prospered, and wrote to the papers to say so. The rollers, he maintained, were light and strong, and there was considerably less friction when they were used. And he challenged the world to disprove his statement. Whereupon an anonymous writer belonging to the rival faction—possibly Joseph Jacob, a coachbuilder who had already written against the system—entered the field, and ventured to suggest that cast iron was exceedingly brittle and not very light. Mr. Sharpe speedily replied. "The principle," he said, and his point is of interest, "upon which rolling carriages are adopted is simply this, That, by the use of them the roads may be made smooth and hard, and by that means, *become part of the mechanism*: for thus the rollers are made to answer all the purposes of light wheels." The anonymous writer appears to have felt the point of this argument, and was forced to retort, quite unworthily, that in any case Mr. Sharpe's rollers were not his own ideas. "No," replied Sharpe, "they were Mr. Daniel Bourn's idea—a very sensible man and good mechanic, and who was also the first contriver of nine-inch broad wheels, who so long as ten or eleven years ago built a waggon on rollers at Leominster where he then lived, and brought it to the Society of Arts and Sciences in the Strand, by whom upon trial it was rejected." The anonymous writer left it at that, but the controversy raged fiercely. It became so highly technical and apparently so interminable that somebody suggested we should all use flying machines and leave the wretched roads to look after themselves.

We may leave the war of the wheels here. The roller-carts were discarded soon afterwards, and M'Adam and his successors rendered for ever such wars unnecessary. But it must not be wholly neglected, and is a tiny chapter by itself in the history of locomotion.

We come to the curiosities.

To this period belongs the present State Coach of Great Britain—that famous "glass-coach" which Londoners had an opportunity of seeing at King George's Coronation. Who built it is not known. Sir William Chambers, "an amateur," as Thrupp is careful to point out, designed it in 1761 for George III. "There is come forth," wrote Walpole to Horace Mann, "a new State Coach which has cost £8000. It is a beautiful object, though crowded with improprieties. Its supports are Tritons, not very well adapted to land carriage, and formed of palm trees, which are as little aquatic as Tritons are terrestrial. The crowd to see it on the opening of Parliament was greater than at the Coronation, and much more damage done."

The ornamentation of the coach, indeed, is a mass of contradictions, but Sir William Chambers did no more than follow tradition. For over a century the principal State Coaches had had Tritons and other queerly inept figures, and Tritons there were in the new coach for King George. Gorgeousness was aimed at, and gorgeousness obtained. There is a detailed contemporary description of this coach which may be given with an account of the expenditure, not quite £8000 as Walpole writes, which it entailed.

> "The carriage is composed of four Tritons, who support the body by cables fastened to the roots of their fins: The two placed on the front of the carriage, bear the driver on their shoulders, and are represented in the action of sounding shells to announce the approach of the monarch of the sea; and those on the back part, carry the imperial fasces, topt with tridents instead of the ancient axes. The driver's footboard is a large scollop shell, supported by a bunch of reeds, and other marine plants. The pole represents a bundle of lances, and the wheels are imitated from those of the

ancient triumphal chariots. The body of the coach is composed of eight palm-trees, which, branching out at the top, sustain the roof. The four angular trees are loaded with trophies, allusive to the victories obtained by *Britain* during the course of the present glorious war. On the center of the roof stand three boys, representing the Genii of *England*, *Scotland*, and *Ireland*, supporting on their heads the Imperial Crown, and holding in their hands the scepter, the sword of state, and ensigns of knighthood. Their bodies are adorned with festoons of laurel, which fall from thence towards the four corners of the roof. The intervals between the palm-trees which form the body of the coach, are filled in the upper parts with plates of glass, and below with pannels adorned with paintings. On the front pannel is represented BRITANNIA seated on a throne, holding in her hand, a staff of liberty, attended by *Religion*, *Justice*, *Wisdom*, *Valour*, *Fortitude*, and *Victory*, presenting her with a garland of laurels. On the back pannel, *Neptune* issuing from his palace, drawn by sea-horses, and attended by the Winds, the Rivers, Tritons, Naids, &c., bringing the tribute of the world to the *British* shore. On one of the doors are represented *Mars*, *Minerva*, and *Mercury*, supporting the Imperial Crown of *Britain*; and on the other, *Industry* and *Ingenuity*, giving a cornucopia to the Genius of *England*. The other four pannels represent the liberal *Arts* and *Sciences* protected; *History* burning the implements of war. The inside of the coach is lined with Crimson Velvet richly embroidered with gold. All the wood work is triple gilt, and all the paintings highly varnished. The harness is of Crimson Velvet, adorned with buckles and other embelishments of silver gilt; and the saddle-cloths are of Blue Velvet, embroidered and fringed with gold."

The account was as follows: —

	£	s.	d.
Coachmaker	1673	15	0
Carver	2500	0	0
Gilder	933	14	0
Painter	315	0	0
Laceman	737	10	7
Chaser	665	4	6
Harnessmaker	385	15	0
Mercer	202	5	10½
Bitt-maker	99	6	6
Millener	31	3	4
Sadler	10	16	6

Woollen-draper	4	3	6
Cover-maker	3	9	6
£7562	4	3½	

Hardly less resplendent was the Lord Mayor's coach which had been built at a cost of over a thousand pounds in 1757, and still performs its duties at stated and regular intervals. It was in 1711 that a Lord Mayor of London had ridden for the last time on horseback in his State procession, this distinction falling to Sir Gilbert Heathcote. Since that date he has been driven in his coach. The 1757 coach was not at first the property of the corporation, but had been built by subscription amongst the aldermen, to whom it belonged until 1778, when the corporation bought it. In that year it had been repaired and repainted—the panels possibly by Cipriani, the heraldic devices by Catton, one of the original members of the Royal Academy and "coach-painter to George III." The Lord Mayor's coach, like many other State coaches of this date, is full of allegorical devices of ornamentation, very plutocratic, very rich, very gorgeous, and incidentally rather more comfortable to drive in than that in which the British Sovereign drives to his Coronation.

Coming to lesser matters, we have mention of a carriage which performed a remarkable feat in 1750.

> "On Wednesday 29," runs a notice of this, "at seven in the morning was decided at *Newmarket* a remarkable wager for 1000 guineas, laid by *Theobald Taaff*, Esq., against the E. of *March* and Lord *Eglington*, who were to provide a four-wheel carriage with a man in it to be drawn by four horses 19 miles in an hour; which was performed in 53 minutes and 27 seconds. The pole was small but lapp'd with fine wire; the perch had a plate underneath, two cords went on each side from the back carriage to the fore carriage, fastened to springs: the harness was of thin leather covered with silk; the seat, for the man to sit on, was of leather straps and covered with velvet; the axles of the wheel were brass, and had tins of oil to drop slowly for an hour. The breechens for the horses were whale-bone; the bars were small wood, straightened with steel-springs, as were most parts of the carriage, but all so light that a man could carry the whole with the harness." Then followed the names of each of the four horses—all had riders—and "lord March's groom sat in the carriage. Two or three other carriages had been made before, but disapproved; and several horses killed in trials—to the expence of 6 or 700*l*."

Now such a carriage—there is a print of it by Bodger—was, of course, little more than a freak. It was a mere skeleton, fragile and entirely useless as a mode of conveyance over the ordinary roads. But the knowledge of those nineteen miles covered easily within the hour must have set people thinking. Such a speed was almost incredible to those accustomed to five or six miles an hour. The carriage itself was the work of Mr. J. Wright, a coachmaker in Longacre, already becoming the home of his brother tradesmen, and it was doubtless exhibited in London. It showed what could be done, and must have opened out agreeable vistas. Twenty miles an hour was something to aim for, and with the war with France concluded, people were able and willing to give rather more attention to the peaceful arts. Amongst other things they showed a desire for strange vehicles. I have mentioned the rolling-carts; there were far queerer carriages, as we shall see, used by the gentry.

"The Carriage Match"

(From a Print by Bodger)

The next curiosity I may speak of was seen in the streets of London during the following year.

> "An odd machine, like an English waggon, drawn by 10 horses, after the *Danish* manner, belonging to Baron Rosencrantz, the new *Danish* envoy, came to his house in *Cleveland Row*, St. James's, from *Harwich*; a coachman drove it and a postilion rode upon the 4th horse."

It suggests rather a primitive type of coach, possibly innocent of springs. What the Baron suffered during his journey through East Anglia must be left to the imagination.

Eight years later, on August 30th, 1758, another strange carriage was seen.

> "This day a remarkable carriage set out from *Aldersgate-street* for Birmingham, from which it arrived on *Thursday* last full of passengers and baggage, without using coomb, or any oily, unctuous, or other liquid matter whatever, to the wheels or axles, its construction being such as to render all such helps useless. The inventor has caused to be engraved on the boxes of the wheels, these words, FRICTION ANNIHILATED, and is very positive that the carriage will continue to go as long and as easy, if not longer and easier, without greasing, than any of the ordinary stage-carriages will do with it: This invention, if really answerable in practice, is perhaps the must useful improvement in mechanicks that this century has produced."

One would like to know who was the inventor of this coach, which, however, did not prosper—I doubt if it performed another journey—for it dropped out of history as suddenly as it had appeared. It would seem that the inventor was a Birmingham man. Possibly he was helped in his scheme by a very extraordinary character who lived and flourished in that town at this time—John Baskerville, successively footman, schoolmaster, graver, japanner, typefounder, and printer—a man whose beautifully printed books have hardly been excelled to this day. Baskerville had made a fortune japanning bread-baskets and the like, and now drove about the country wonderfully dressed in a coach apparently of his own design—he was a man who had to do everything for himself, and being of somewhat eccentric disposition, never did anything like anybody else—and his coach, like his house and his printing and his religious opinions, was like nothing in the world. He had a considerable idea of his own importance, and his coach was a reflection of his character. With its wonderful arms—the real Baskerville arms, to which the printer had no right whatever—it was standing until quite recently in an old barn in a field at Manton. It was thus described fifty years ago:—

> "The body hangs by double straps, from the coachman's seat under the carriage, to which they are fastened, to the frame behind.... It could be either closed or open, and when open the leather top was rolled back upon crossed straps hung from the coachman's seat, and hooks secured to the front part of the body. The

whole framework of the carriage has been elaborately carved and gilt, and the panels painted with what appears to be a brownish green, with flowers and vases, rock and shell-work, among which were numerous figures of boys and emblems. In the centre panel on each side were the arms, on the side panel the crest...."

None of the panels were identical, but all had been decorated by his workmen. "The pattern-cart of his trade," Hutton, the Birmingham historian, calls this curiosity, which was once familiar to every village in the Midlands, and his daughter, Catherine Hutton, could remember the printer, "in his gold-laced waistcoat, and his painted chariot, each panel a picture, fresh from his own manufactory of japanned tea-boards."

A most extraordinary conveyance appeared in London in 1771—this being "Mr. Moore's new-invented Coal-carriage," the wheels of which were no less than fifteen feet high.[43] A great concourse of people followed it through the streets, and no doubt applauded its ability to draw two caldrons and two sacks of coal, using only two horses abreast, "with more ease and expedition than the common carts do one caldron with three horses at length." Unfortunately I have not been able to discover a print of this monstrous vehicle, which, like so many of the other mid-century freaks, disappeared almost at once.

To this period also belongs that wondrous *phaeton*, which in a few years threatened to become so lofty as to suggest to some ingenious artist the possibility of applying to it some pantograph arrangement whereby its seat could be raised or lowered at will. This print, called The *New Fashioned Phaeton—Sic itur ad Astra*, was published in 1776, a curious mezzotint showing a lady of fashion stepping out of a first-floor window into the seat of a phaeton which has been raised to the required height. The phaetons, indeed, seem to have been built high since their invention, and the importance of this feature must not be overlooked, when one remembers that almost every carriage, both English and foreign, was hung enormously high in the last years of the century, nine or ten steps being sometimes necessary to get inside.

Exactly when or where the phaeton was first made I cannot determine, but, like the *landau*, which has generally, though incorrectly, been considered to have been first built in 1757, it is mentioned so early as 1747 in the poem quoted at the end of the last chapter. That it was already popular with the fashionable people is shown by Tom Warton's poem, *The Phaeton and the One Horse Chair*, which was first published in the *Gentleman's Magazine* for December, 1759. This is worth quoting in its entirety:—

"At *Blagrave's* once upon a time,
There stood a phaeton sublime:

[43] This was the Mr. Francis Moore, of Cheapside, who in 1786 and 1790 obtained patents for two two-wheeled carriages. The second of these bore considerable resemblance to the hansom-cab of a later date. It had enormous wheels—higher, indeed, than the body of the carriage—and the driver sat on a small box-seat in front and at a level with the top of the roof. The door was at the back.

Unsully'd by the dusty road
Its wheels with recent crimson glow'd;
Its sides display'd a dazzling hue,
Its harness tight, its lining new:
No scheme-enamoured youth, I ween,
Survey'd the gaily deck'd machine,
But fondly long'd to seize the reins,
And whirl o'er *Campsfield's* tempting plains.
Mean time it chanc'd, that hard at hand
A one-horse chair had took its stand;
When thus our vehicle begun
To sneer the luckless chair and one.
'How could my master place me here
Within thy vulgar atmosphere?
From classic ground pray shift thy station,
Thou scorn of *Oxford* education!
Your homely make, believe me, man,
Is quite upon the Gothic plan;
And you, and all your clumsey kind,
For lowest purposes design'd:
Fit only with a one ey'd mare,
To drag, for benefit of air,
The country parson's pregnant wife,
Thou friend of dull domestic life,
Or, with his maid and aunt, to school,
To carry *Dicky*, on a stool.
Or, haply to some christ'ning gay,
A brace of godmothers convey.—
Or, when blest *Saturday* prepares
For *London* tradesmen rest from cares,
'Tis thine, o'er turnpikes newly made,
When timely show'rs the dust have laid,
To bear some alderman serene
To *fragrant Hampstead's sylvan* scene.
Nor higher scarce thy merit rises
Among the polish'd dons of *Isis*.
Hir'd for a solitary crown,
Canst thou to *schemes* invite the *Gown*?
Go, tempt some prig, pretending taste,
With hat new cock'd and newly lac'd,
O'er mutton chops, and scanty wine,
At humble *Dorchester* to dine!
Mean time remember, lifeless drone!
I carry *Bucks* and *Bloods* alone.
And oh! when 'er the weather's friendly,
What inn at *Wallingford* or *Henley*,

But still my vast importance feels,
And gladly greets my entring wheels.
And think, obedient to the throng,
How yon gay streets we sneak along:
While all with envious wonder view
The corner turn'd so *quick* and *true*.'
To check an upstart's empty pride,
Thus sage the one horse chair reply'd.
'Pray, when the consequence is weigh'd
What's all your spirit and parade?
From mirth to grief what sad transitions,
To broken bones—and impositions!
Or if no bones are broke, what's worse,
Your *schemes* make work for *Glass* and *Nourse*.
On us pray spare your keen reproaches,
From one-horse chairs men rise to coaches;
If calm discretion's steadfast hand,
With cautious skill the reins command,
From me fain health's fresh mountain springs,
O'er me soft snugness spreads her wings:
And innocence reflects her ray
To gild my calm sequester'd way;
E'en kings might quit their state to share
Contentment and a one horse chair.—
What though, o'er yonder echoing street,
Your rapid wheels resound so sweet,
Shall *Isis'* sons thus vainly prize
A rattle of a *larger size*?'
Blagrave, who during the dispute,
Stood in a corner, snug and mute,
Surpriz'd no doubt, in lofty verse,
To hear his carriages converse,
With solemn care, o'er *Oxford* ale,
To me disclos'd this wondrous tale.

MORAL

"Things may be useful if obscure;
The pace that's slow is often sure;
When empty pageantries we prize,
We raise but dust to blind our eyes.
The Golden Mean can best bestow
Safety for unsubstantial *Show*."

From this poem it is possible to understand that this new-fangled carriage was used rather as a toy than anything else. That it was dangerous clearly appears, and it was this very danger which must have contributed not a little to its popularity. It was driven at a very great rate, and with a recklessness that excited the anger of the commoner folk—unless, as was often the case, it excited their admiration instead. The phaeton was the most sporting carriage you could have. It lent itself to the idea of racing, and there was always the chance that an accident might be fatal—an allurement in itself. And so in a very few years there was hardly a fashionable young gentleman in London who did not possess one of these carriages and drive about, insolently staring down from his enormously high seat on to the heads of the crowds below.

"Phaetona, or Modern Female Taste," 1776

Experiments, too, were being made with them. The position of the body was gradually brought forward until it was directly over the front axle. In 1766 "the Hon. Sir Francis Blake Delavel, Knight of the Bath," was experimenting with a "new-invented phaeton the other side of Westminster Bridge, where he put his horses in a full gallop, and in a moment, by pulling a string, the horses galloped off and left him in the carriage, which stood still." Sir Francis was apparently working at some contrivance to be used in case the horses chose to run away—a common occurrence, no doubt, and apt to be far more dangerous to the driver than would be the case with other carriages, for the body of these early phaetons was slung high above the undercarriage by the most delicate supports, which bent and creaked and were obviously unfitted to bear any great strain. The body itself must have resembled that of the curious chaises which were still to be seen at this time in France and Italy—just a small chair varnished and sometimes painted, fixed to four thin and often carved and curled posts, which as often as not rose merely from the shafts, there being no springs of any kind. The shafts were very long, and the common practice seems to have been to drive two horses tandem, with, no doubt, a postilion on the leader. The phaeton was probably slimmer than these equally curious vehicles, and much higher, and their ability to turn corners with ease may be deduced from the lines just quoted.

"Sir Gregory Gig"

(From a Print by Bunbury, 1782)

A phaeton built for a lady is shown in a print published in 1776, called *Phaetona; or Modern Female Taste*. Here the carriage has a very small body, hung very high on large wheels, the undercarriage being abnormally long in consequence. The two horses which draw it are very undersized—another peculiarity possibly demanded by contemporary fashion.

Two years later the scandalous *Town and Country Magazine* published a short and probably true tale called *The Rival Phaetons*, which shows to what lengths, or, rather, what heights the Bucks of the time would go.

"Lord M— —," it runs, "emulous of shining in the most elevated sphere, first drove a phaeton seven feet from the ground: Sir John L[ade] immediately made an addition of a supernumerary travelling case to his, and raised it six inches higher. Lord M— — applied immediately to his coachmaker in Liquor-pond-street for two travelling cases, with which he speedily drove about the streets for the entertainment of the public. Sir John L[ade] was stung to the quick; and Lord M— — 's round hat was now a mere pigmy to his. His Lordship, happy at rival inventions, immediately added two more horses to his triumphal car, and drove four for expedition, from Grosvenor Square to Gray's-inn-lane. 'Now, my Lad,' said he, 'I have you;' but how vain are the boastings of mankind? The knight appeared the very next day with a phaeton and six in Holborn. 'Zounds,' said his lordship, 'this is too much! what shall I do?—how can I match my four with two more? No credit at my banker's—in arrears with my horse-dealer—I am at my wit's end. John, I shall not take an airing in Smithfield to-day; I'll give my horses some rest—they were hard worked over the stones yesterday.' Here the contest now lies—its importance must be obvious to every beholder—his lordship has not slept these three nights, and it is imagined he will at length be obliged to take the hint from Colman's prologue to the Suicide, and preposterous as it may appear, add a fifth wheel to his phaeton. Sir John is greatly elated, and may literally be said to be in very high spirits upon his temporary triumph."

Writing to Mann in June, 1755, Walpole, after regretting the absence of social news in England, mentions the latest Paris fashion. "All the news from France," he says, "is that a new madness reigns there, as strong as that of Pantins was. This is *la fureur des cabriolets, Anglicè*, one-horse chairs, a mode introduced by Mr. [Josiah] Child; they not only universally go in them, but wear them; that is, everything is to be *en cabriolet*; the men paint them on their waistcoats and have them embroidered for clocks to their stockings; and the women, who have gone all the winter without anything on their heads, are now muffled up in great caps [calash hoods] with round sides, in the form of, and scarce less than the wheels of chaises."

"The cabriolet head-dress," says Wright,[44] "was soon improved into *post-chaises, chairs-and-chairmen*, and even *broad-waggons*." So we have *A Modern Morning*, published in 1757:—

> "Then Caelia to her toilet goes,
> Attended by some favourite beaux.
>
> 'Nelly! why, where's the creature fled?
> Put my post-chaise upon my head.'
> 'Your *chair-and-chairman*, ma'am, is brought.'
> 'Stupid! the creature has no thought!'
> 'And, ma'am, the milliner is come,
> She's brought the *broad-wheel'd waggon* home.'"

In which structures Caelia sallies forth.

These cabriolets rivalled the phaetons as fashionable carriages, and indeed as the *new gigs* came to resemble them in every point save the number of the wheels. There is a print by Colley, dated 1781, showing one of these new gigs. The small chair, very high, holding two people, is supported by long curved supports, which in themselves of course acted as springs of a kind. Two horses are being driven tandem, with a postilion driving the leader. Another print, by Bunbury, called *Sir*

---

[44] *Caricature History of the Georges*, Thomas Wright. London, n.d.

Gregory Gigg, shows a young man driving a pair of horses abreast. He is seated in a still smaller, and slightly lower, chair. This was a *curricle* rather than a cabriolet, and it was such a carriage which the braggart sportsman, John Thorpe, describes to Catharine in *Northanger Abbey*: "Curricle-hung, you see, seat, trunk, sword-case, splashing board, lamps, silver-moulding, all, you see, complete; the ironwork as good as new, or better. He [the first owner] asked fifty guineas; I closed with him, threw the money down, and the carriage was mine." The shape of these curricles is well seen in Bunbury's drawing.

A glance at the newspaper advertisements of the day will afford an insight into the various carriages in use. So, for instance, in 1767 we have:—

> An exceeding good Post chariot, the Box to take off.

> A neet genteel Single Horse Chaise, painted green, and hung upon Steel Springs.

> An exceeding fine black gelding that goes well in an Italian Chair, with a Tail.

> A very neat fashionable Chaise.

> A very good second-hand Phaeton Chaise, that goes either with one horse or two, with Shafts, Poles, and Harness suitable, Steel Springs, and Iron Axletrees. Also a good second-hand Landau, which alters occasionally into a Phaeton, steel springs and Iron Axletrees to the Carriage.

The landau, by the way, was a recent invention (though made, as we have seen, before 1757) which may be dismissed with the observation that it was a coach made to open when required.

And put up for auction together on one occasion were:—

> A green windsor chair,
> A good Post-Coach,
> A Post Landau,
> A very neat Italian Chair,
> 3 old Chariots,
> 4 Post-Chaises, and
> 3 single Chaises.

So run these advertisements, with scraps of information interspersed and little puffs of the advertiser on every other line. What the windsor chair was I have not been able to discover; but it is to be noticed both that Italian chairs (or chaises) were apparently popular, and that the English-built carriages were being constructed on rather a loftier scale. The curious reason for this will appear in the next chapter.

Meanwhile I may conclude by drawing attention to two other advertisements of a curious nature.

The first of these deals with a hackney coachman who had refused to carry a fare. The second, which I do not think has been reprinted since it originally appeared in 1767, shows the dangers to which travellers were still liable.

From the time when the dramatist Congreve had been appointed a Commissioner for Licensing Hackney Coaches (1695) there had been frequent legislation with regard to these hackney coaches. At this time there were stringent regulations, some of which are still in force, with regard to the taking up of passengers. It was the refusal of a coachman to drive a gentleman who had hailed him that led to the following pitiful notice:—

"Whereas I William Ford, late driver of an hackney coach, No. 694, did refuse to carry a gentleman, and did also grosly abuse him; for this I was fined thirty shillings by the Commissioners. I then most wickedly and falsely swore an assault against, and had the same gentleman carried before Sir John Fielding, who discharged the warrant. For this false imprisonment, I had a prosecution commenced against me, and though I made frequent application for pardon, I could not obtain it until the expence amounted to a sum which has almost ruined me, and which I have paid. I therefore voluntarily [?] insert this as a caution to other hackney coachmen, who well know that it is from the hope of forgiveness, which they too often meet, that they venture so daringly to abuse and insult their fare.

"WILLIAM X FORD
"His mark."

It was this same Sir John Fielding, the blind magistrate, who inserted, some little time afterwards, the following warning to travellers and others:—

"To the Stage Coachmen, Carriers, Book-keepers,
"Tradesmen in general, and others.

"Public Office, Bow Street, September 24, 1767.

"A most necessary caution at this season of the year.

"The remainder of that Gang of unhappy wretches, who live in Idleness and subsist on Plunder, and who make it their particular Business, from this Time to the End of Winter to cut off Trunks from behind Post Chaises, to steal Goods out of Waggons, from the Baskets of Stage-Coaches, Boots of Hackney Coaches, and out of Carts which carry Goods to and from Inns, &c. (though but few in Number) having already begun to wait in the Dusk of the Evenings, at the different Avenues leading to Town, and at several Inns, &c., for the above Purposes; 'tis hoped that an Attention to the following Observation, may be the Means of preserving much Property, which when once lost by these Means, is difficult to recover, or the Offenders to be detected.

"1. Those who cannot conveniently fasten their Luggage before them in Post Chaises, should take care to secure it behind with a small Chain instead of a Rope or Strip, and to place the Padlock that fastens it out of Sight or Reach; and those who have

Servants to attend them, should direct them to keep close to the Carriage as they come to London, for these Plunderers extend themselves for fifteen Miles out of Town to the very Inns themselves in London, and are ready in an amazing Manner to take Advantage of the least Neglect.

"2. As it is common for Persons on their Arrival in Town to take a Hackney Coach when they come on the Stones, in the Boot of which they generally deposit their Luggage, they should be cautious never to send the Coachman from his Box, to make an Enquiry, &c. for if he be absent a Minute his Fare will be in great danger of losing his Property, by some of the above Offenders, who attend at the Inns at the Entrance of the Town, in order to follow Hackney Coaches to the Places where they set down or stop, to watch an Opportunity to plunder.

"3. Nothing can secure the Goods in Waggons, or the Baskets of Stage Coaches, but the Care of the Drivers, who should have them watched both on and off the Stones, and the Proprietors of the several Road Waggons should have a Man at least on Purpose to guard them five or ten Miles out of Town, a step which is absolutely necessary.

"J. Fielding."

Also, of course, there were the highwaymen.

CHAPTER THE EIGHTH
THE AGE OF TRANSITION

"So down thy hill, romantic Ashbourne, glides
The Derby Dilly, carrying three insides.
One in each corner sits and lolls at ease,
With folded arms, propt back, and outstretched knees;
While the press'd Bodkin, pinch'd and squeezed to death,
Sweats in the mid-most place, and scolds, and pants for breath."

Canning.

"IN the year 1790," wrote Mr. William Felton in an account of the carriages of his day, "the art of Coach-building had been in a gradual state of improvement for half a century past, and had now arrived at a very high degree of perfection, with respect to the beauty, strength, and elegance of our English carriages." And the most cursory glance at his carefully compiled, if technical book, is evidence enough of the truth of his statement. At this time, indeed, the old flamboyant ornamentation had all but disappeared from the carriages, which were in process of taking on the appearance they largely retain to this day. Most vehicles, it is true, were still hung far higher than those of the nineteenth century — a fact due to the curious, though mistaken, belief, "that a high and short load possessed some mysterious property which made it easier to draw than a long one," but new principles were being adopted as the result of careful experiments. Prizes were offered by learned societies, and won. Men like Dr. Lovell Edgeworth, who had been experimenting so early as 1768, and had shown that springs — then but little understood — were at least as advantageous to the horses as to the passengers, were at work. But it was only in 1804, when Mr. Obadiah Elliott produced his patent elliptic springs, which rendered unnecessary the old heavy perch, that a definite period in the art of coach-building was clearly marked. Thenceforth the older, cumbrous machines disappeared from the roads and made way for the lighter and more comfortable carriages which were to be seen at the time of Queen Victoria's accession.

The question of the roads, too, was receiving the attention of experts. Anstice and Edgeworth published the results of their investigations, but were both completely overshadowed by James M'Adam, who about 1810 started those metal roads which have proved so wonderfully successful. Before his time gravel and the like had formed the basis of road-material; M'Adam used granite and other allied substances, and produced such a surface as had not been seen since the Romans had constructed their vast highways hundreds of years before.

Methods of travelling, moreover, were altering. The stage-coaches, useful though they were, disappeared before Palmer's mail-coaches, which held their supremacy until the era of steam revolutionised locomotion. Post-chaises were still in favour, and less dangerous than of old. Incidentally, the highwaymen were taking to less romantic pursuits. And what is true of England was also in a great measure true of Europe as a whole. North America, too, at this period was providing herself with coachbuilders, who produced distinctive vehicles peculiarly adapted to the conditions of that country.

It was, in fact, a transition period.

We may consider in the first place such types of carriages as already existed. There is a whole catalogue of them, and only one of the older carriages is conspicuously absent. This was the calash—"now almost obsolete for any purpose," comments Leigh Hunt, and indeed there is hardly a reference to it. But the others still survived, and one characteristic is immediately noticeable; the wheels of almost every sort of carriage at this time were enormously large. Consequently the carriages were generally very long. Crane-neck perches were still used, and what was called an upright spring. A coach of this period, belonging to the museum at South Kensington, is now exhibited in Edinburgh. It was built for the Lord Chancellor of Ireland. It has "a large body," says Thrupp, describing it, "with deep panels, flat-sided, longer on the roof than at the elbow, with windows in the upper quarters; the carriage with two crane perches (easily seen in the accompanying photograph), Berlin fashion, whip springs, and very high wheels. There is no footboard, whilst a hammercloth for the footman is raised upon scroll ironwork, very well made." Napoleon's state coach, built at the time of his second marriage, and preserved at Vienna along with a chariot and barouche, is of a somewhat similar pattern. His travelling coach, with all its household contrivances, is now at Madame Tussaud's exhibition, and must be familiar to all Londoners. Two Spanish coaches of the period are also to be seen at Madrid.

The Lord Chancellor's coach was of course an exceptional carriage, and Mr. Felton is careful to give details of such lesser coaches as were being made. These he catalogues as a plain coach, a neat ornamental town coach, a landau, a travelling coach, an elegant crane-neck coach, and a vis-à-vis, which last, he says, "is seldom used by any other than persons of high character and fashion." And, indeed, this particular carriage is to be seen in numerous plates and caricatures of the time.

Coming to the chariots and post-chaises, there is a good example of an English carriage of the kind at South Kensington. This apparently belonged to George III. The photograph gives but a poor idea of the great size of the original. The wheels are taller than an average man, and the length of the carriage is prodigious. The single window on either side is small, the panels are deep, and there is a small platform at the back of the body to carry luggage. A footboard still remains with supports for the driver's seat that has disappeared.

George III's Posting Chariot

(At South Kensington)

It was in such a chariot, though even larger than George III's, that the unhappy King and Queen of France attempted to escape from Paris—that "miserable new Berline," as Carlyle calls it, which was the very last carriage to be used for such a purpose.

"On Monday night, the Twentieth of June, 1791," runs Carlyle's own wonderful account, "about eleven o'clock, there is many a hackney-coach and glass-coach (*carrosse de remise*), still rumbling, or at rest, on the streets of Paris." Into one of these glass-coaches steps "a hooded Dame with two hooded Children, a thickset Individual, in round hat and peruke." The coachman is Fersen himself.

"Dust shall not stick to the hoofs of Fersen: crack! crack! the Glass-coach rattles, and every soul breathes lighter. But is Fersen on the right road? Northeastward, to the Barrier of Saint Martin and Metz Highway, thither were we bound; and lo, he drives right Northward! The royal Individual, in round hat and peruke, sits astonished; but right or wrong, there is no remedy. Crack, crack, we go incessant, through the slumbering City. Seldom, since Paris rose out of mud, or the Longhaired Kings went in Bullock-carts, was there such a drive. Mortals on each hand of you, close by, stretched out horizontal, dormant; and we alive and quaking! Crack, crack, through the Rue de la Chaussée d'Antin,—these windows, all silent, of Number 42, were Mirabeau's. Towards the Barrier not of Saint-Martin, but of Clichy on the utmost North! Patience, ye royal Individuals; Fersen understands what he is about. Passing up the Rue de Clichy, he alights for one moment at Madame Sullivan's: 'Did Count Fersen's Coachman get the Baroness de Korff's new Berline?'—'Gone with it an hour-and-half ago' grumbles responsive but drowsy porter. '*C'est bien.*' Yes, it is well;—though had but such hour-and-half been lost, it were still better. Forth therefore, O Fersen, fast, by the Barrier de Clichy; then Eastward along the Outer Boulevard, what horses and whipcord can do!

"Thus Fersen drives, through the ambrosial night. Sleeping Paris is now all on the right-hand of him; silent except for some snoring hum: and now he is Eastward as far as the Barrier de Saint-Martin; looking earnestly for Baroness de Korff's Berline. This Heaven's Berline he at length does descry, drawn up with its six Horses, his own German Coachman waiting on the box.... The august Glass-Coach fare, six Insides, hastily packs itself into the new Berline; two Bodyguard Couriers behind. The Glass-coach itself is turned adrift, its head towards the City; to wander whither it lists,—and be found next morning in a ditch. But Fersen is on the new box, with its brave new hammer-cloths; flourishing his whip; he bolts forward towards Bondy. There a third and final Bodyguard Courier of ours ought surely to be, with post-horses already ordered. There likewise ought that purchased Chaise, with the two waiting-maids and their band-boxes, to be; whom also her Majesty could not travel without....

"Once more by Heaven's blessing, it is all well. Here is the sleeping Hamlet of Bondy; Chaise with Waiting-women; horse all ready, and postilions with their churn-boots, impatient in the dewy dawn. Brief harnessing done, the postilions with their churn-boots vault into the saddles; brandish circularly their little noisy whips...."

The Lord Chancellor of Ireland's Coach

(Now in Edinburgh)

"But scouts, all this while, and aides-de-camp, have flown forth faster than the leathern Diligences...."

The grand new Berline has been seen in the Wood of Bondy.

"Miserable new Berline!" apostrophises Carlyle. "Why could not Royalty go in some old Berline similar to that of other men? Flying for life, one does not stickle about his vehicle. Monsieur, in a commonplace travelling-carriage, is off Northwards; Madame, his Princess, in another, with variation of route; they cross one another while changing horses, without look of recognition; and reach Flanders, no man questioning them....

"All runs along, unmolested, speedy, except only the new Berline. Huge leathern vehicle:—huge Argosy, let us say, or Acapulco-ship; with its heavy stern-boat of Chaise-and-pair; with its three yellow Pilot-boats of mounted Bodyguard Couriers, rocking aimless round it and ahead of it, to bewilder, not to guide! It lumbers along lurchingly with stress, at a snail's pace; noted of all the world."

It has indeed been seen, and soldiers rush after it, and the huge Berline is brought back to Paris in what was surely the most terrible procession ever witnessed....

The Korff Berline was probably not built so high as some of the English posting chariots of the time. The perch of these was often more than four feet from the ground. According to Felton you could buy a plain post-chaise for £93, or a neat town chariot for £91. Or you might have a landaulet, a demi-landau, or a sulky, which at this time was "a light carriage built exactly in the form of a post-chaise, chariot, or demi-landau," and like the vis-à-vis was "contracted on the seat, so that only one person can sit thereon, and is called a sulky from the proprietor's desire of riding alone." The landaulet was to the landau as the chariot was to the coach. It was simply a chariot made to open. The hood was of "greasy harness leather, disagreeable to the touch or smell, and continually needing oil and blacking" rubbed into it to keep it supple and black.

Then there was the phaeton, which had lost none of its popularity, and was built as lofty as ever.

"The handsomest mixture of danger with dignity," wrote Leigh Hunt, "in the shape of a carriage, was the tall phaeton with its yellow wings. We remember looking up to it with respect in our childhood, partly for its loftiness, partly for its name, and partly for the show it makes in the prints to novels of the period. The most gallant figure which modern driving ever cut was in the person of a late Duke of Hamilton; of whom we have read or heard somewhere, that he used to dash round the streets of Rome, with his horses panting, and his hounds barking about his phaeton, to the equal fright and admiration of the Masters of the World, who were accustomed to witness nothing higher than a lumbering old coach, or a cardinal on a mule."

But far more conspicuous a figure than this Duke of Hamilton was Colonel (Tommy) Onslow, afterwards Lord Cranley, of whom there is a caricature by Gillray, with the following once famous lines:—

"What can little T. O. do?
Why drive a phaeton and two.
Can little T. O. do no more?
Yes, drive a phaeton and four!"

The Colonel, however, was surpassed, as we have seen, by Sir John Lade, who drove six greys. George IV, when Prince of Wales, was satisfied with a pair, but his horses were "caparisoned with blue harness stitched with red," their manes "being plaited with scarlet ribbons, while they wore plumes of feathers on their heads."

The structure of these phaetons differed. Gillray's picture shows the body hung midway between the two axles, though he may not have troubled to be exact in this respect. The commonest form was the *perch-high phaeton*, in which the body was hung directly over the front axle, the hind wheels being much larger than those in front, and the bottom of the body being five feet from the ground. Others were less lofty. In the *one-horse phaeton* the body was hung over the back axle with "grasshopper" springs, and "was joined to the forecarriage, which was without springs, by wooden stays"—a very different carriage. This in time led to the *pony phaeton* used by George IV in 1824. Here all idea of great height had been abandoned so as to allow His Majesty to enter his carriage without the fatigue of climbing several steps. Queen Victoria's pony phaeton was a similar vehicle, and indeed it was from such a carriage that the *victoria* was evolved at a rather later date.

"What connexion there could be," wrote Bridges Adams some forty years later in a passage not altogether devoid of epithets, "between this vehicle and the fabled car of the Sun-God, to obtain for it such a title, it is difficult to conceive.... The vehicle looked like a mechanical illustration of the play of *Much Ado about Nothing*. It was a contrivance to make an enormously high and dangerous seat for two persons, inconvenient to drive from, and at the same time to consume as much material and mix as many unsightly and inharmonious lines as possible. The framework of the carriage was constructed with two iron perches, the outline of which was hideously ugly; but the camel-like hump had at least the mechanical advantage of permitting a higher fore wheel than could otherwise be used. The shape of the body was as though the rudest possible form capable of affording a seat had been put together. An ungraceful form of upright pillar or standard was first selected, into which was framed a horizontal ugly curve for a seat, connected at the top by an ungainly-looking elbow, and a formal serpentine curve behind, from which was projected like an excrescence an ugly leathern box called a sword-case. The front of the upright pillar was continued into a most formal curve, and from its point rose an ungraceful bracket, to support a footboard, on the extreme edge of which was coiled an ugly piece of leather called an apron. The construction of the body was such that it could not possibly hold together by the strength of its own framing; and to remedy this, a curved iron stay was introduced in the worst possible taste.... The fore springs rather resembled the flourishing strokes made by a schoolmaster, when heading a copy-book or Christmas piece, than any legitimate mechanical contrivance; and the motion must have been detestable, rendering the act of driving difficult, and lessening the power of the drivers over their horses. The servant's seat behind" —not always present— "placed on curved blocks without any springs, completed this extraordinary-looking vehicle. To sit on such a seat, when the horses were going at much speed, would require as much skill as is evinced by a rope-dancer at the theatre."

Which shows that in 1837, at any rate, people's ideas had undergone a considerable change with regard to a really fashionable equipage.

The only other four-wheeled vehicle I need mention here was the *sociable* which, according to Felton, was "merely a phaeton with a double or treble body." It was made with or without doors, and with or without a driving seat. A good example of this carriage is shown in Gillray's print *The Middlesex Election of 1804.*

Coming to the two-wheeled vehicles, the chief of these were the *curricle*, the *gig* or *chaise*, and the *whiskey*. As a general rule it may be taken that when a gig had two horses it was called a curricle, and when there was only one, a chaise. In the Prince Regent's time the curricle was "the most stylish of all conveyances." In shape nearly all these gigs were identical, though one reads that the notorious "Romeo" Coates drove in one whose body was shaped like a shell.[45] They were of various heights, a particularly lofty one being known in Ireland as the *suicide gig*. The *caned whiskey* was a gig whose body, "fixed upon the shafts—which again were connected with the long horizontal springs by scroll irons," had a movable hood. The *Rib Chair* was similar to the whiskey, but without springs. It is really only possible to differentiate properly between these light carriages and the other hybrids, so soon to appear, by means of prints and photographs. To the non-technical mind they are almost identical with each other.

"The prettiest of these vehicles," Leigh Hunt writes, after confessing that he has no ambition to drive tandem, as was so often done, or to run into danger with a phaeton, "is the curricle, which

---

[45] "The shape of the body," says Bridges Adams, describing Coates's carriage, "was that of a classic sea-god's car, and it was constructed in copper. This vehicle was very beautiful in its outline, though disfigured by the absurdity of its ornamental work." When Coates had a fall, Horace Smith, of *Rejected Addresses* fame, seized the occasion to write a mock condoling poem.

is also the safest. There is something worth looking at in a pair of horses, with that sparkling pole of steel laid across them. It is like a bar of music, comprising their harmonious course. But to us, even gigs are but a sort of unsuccessful gentility. The driver, to all intents and purposes, had better be on the horse."

I need say very little of the public carriages. There is, however, one point in connection with the later stage-coaches which bears upon the question that was only solved by Obadiah Elliott in 1804. On September 20, 1770, according to the *Annual Register*, there was an accident to one of them which was growing increasingly common.

"It were greatly to be wished," runs this account, "the stage coaches were put under some regulations as to the number of persons and quantity of baggage. Thirty-four persons were in and about the Hertford Coach this day when it broke down by one of the braces giving way."

No wonder it broke down! It is interesting to note, however, that even the more humane stage-coachmen, so far from objecting, as you might imagine they would have done, to such overcrowding, actively encouraged it and for a very odd reason. At this time springs of a kind were being applied to the coaches, which consequently travelled with greater ease than before, but the coaches themselves happened also to be built very high, like all other vehicles, and nothing could convince the silly coachmen that the easy running was not due to a heavy load being applied to the top of a high carriage. It became necessary, therefore, to pass legislation, which was accordingly done in 1785 and again in 1790, restricting the number of passengers allowed.

At this time, too, Mr. John Palmer's first diligence, or mail-coach, had appeared as a quick and cheap method of carrying letters, and these mail-coaches very rapidly took the fancy of passengers. Palmer, however, was a man with great powers of organisation, and before the new century had dawned, had his coaches running upon every high road in the country.[46]

"The mail coaches," wrote a French nobleman after visiting this country at the beginning of the new century, "afford means of travelling with great celerity into all parts of England. They are Berlins, firm and light, holding four persons; they carry only letters, and do not take charge of any luggage. They are drawn by four horses, and driven by one coachman; they travel never less than seven to eight miles an hour."

One or two particular inventions may also be noted. This same nobleman, continuing his account, says:—

"Stage Coaches are very numerous, they are kept in every City, and even in small towns; all these carriages have small wheels, and hold six persons, without reckoning the outside pas-

[46] For a detailed account of these mail-coaches the reader is referred to Mr. Charles Harper's book, *Stage Coach and Mail in the Days of Yore*.

sengers. About twenty years ago a carriage was invented in the form of a gondola; it is long, and will hold sixteen persons sitting face to face; the door is behind, and this plan ought to be generally adopted, as the only means of escaping a great danger when the horses run away. What adds to the singularity of these carriages is, that they have eight wheels; thus dividing equally the weight, they are less liable to be overturned, or cut up the roads; they are, besides, very low and easy.

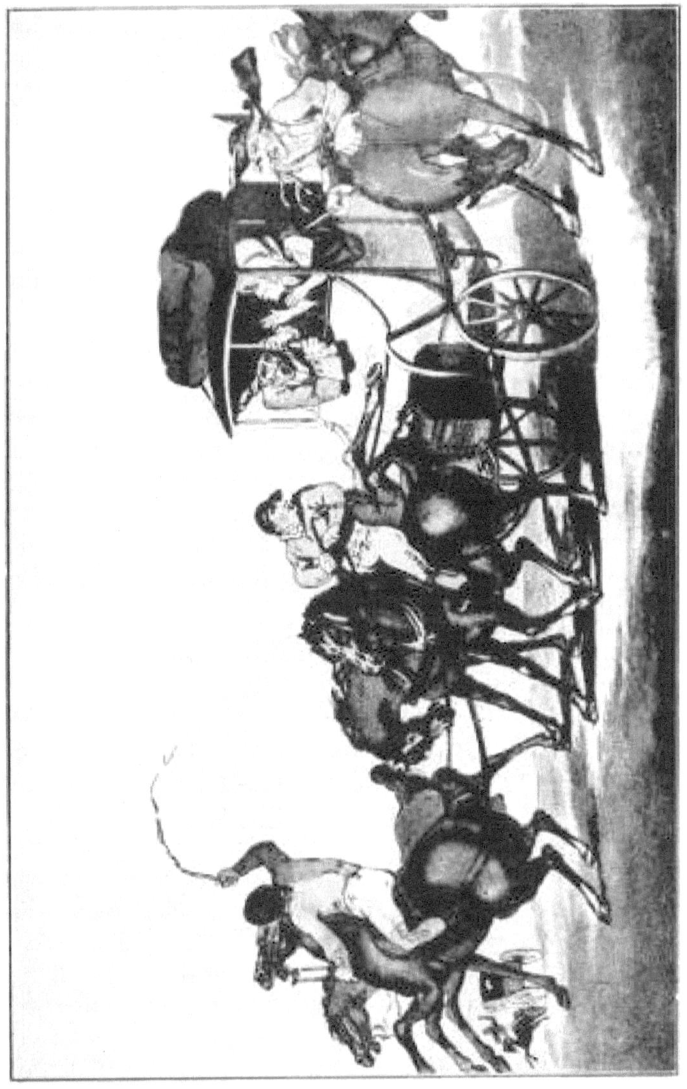

"English Travelling, or the First Stage from Dover"
(From a Drawing by Rowlandson, 1792)

"When these long coaches first appeared at Southampton, a City much frequented in summer by rich inhabitants of London, who go there to enjoy sea bathing, they had (as every new thing has) a great run, so that it was nearly impossible to get a place in them.

"One of the principal Innkeepers, jealous of this success, set up another, and, to obtain the preference, he reduced the fare to half-price, at that time a guinea. In order to defeat this manœuvre, the first proprietor made a still greater reduction, so that, at last, the receipts did not cover the expenses. But the two rivals did not stop here; for one of them announced that he would take nothing of gentlemen who might honour him by choosing his Coach, but he would beg them to accept a bottle of Port before their departure."

But not even such a temptation seems to have made these long coaches a success.

The other innovation, though properly belonging to a slightly later date, was the patent coach invented by the Reverend William Milton. He explained his coach in a letter to Sir John Sinclair.[47]

"Permit me, Sir, to explain, in a few words, the nature of my invention.—In a stage-coach, an overturn is rendered much less likely to happen, by placing as much as possible of the heavy luggage of each journey, in a luggage-box below the body of the carriage; the body not being higher than usual. This brings down the centre of gravity of the total coach and load (a point which at present, at every inequality of the road and change of quarter, vacillates most dangerously), it brings it down to a place of great comparative safety.

"To prevent the fatal and disastrous consequences of breaking down, there are placed, at the sides or corners of this luggage-box, small strong idle wheels, with their periphery below its floor; ready, in case of a wheel coming off or breaking, or an axle-tree failing, to catch the falling carriage, and instantly to continue its previous velocity; thereby preventing that sudden stop to rapid motion, which at present constantly attends the breaking down, and which has so frequently proved fatal to the coachman and outside passengers.—The bottom of this luggage-box is meant to be about twelve or thirteen inches from the ground, and the idle wheels seven, six, or five. If at a less distance still, no inconvenience will result; for when either of them takes over an obstacle in the road, it instantly, and during the need, discharges its respective active wheel from the ground, and works in its stead."

Several coaches were built to Mr. Milton's specifications, but like so many other

[47] *The Danger of Travelling in Stage-Coaches; and a Remedy Proposed to the Consideration of the Public,* by the Rev. William Milton, A.M., Vicar of Heckfield, Hants. Reading, 1810.

patent coaches they were speedily forgotten.

It is only necessary to add here that about 1800 "outside passengers were first enabled to ride on the roofs of coaches without incurring the imminent hazard of being thrown off whenever their vigilance and their anxious grip relaxed." For it was then, says Mr. Harper, "that fore and hind boots, framed to the body of the coach, became general, thus affording foothold to the outsides. Mail coaches were not the cause of this change, for they originally carried no passengers on the roof. We cannot fix the exact date of this improvement," he adds, "and may suppose that in common with every other innovation, it was gradual, and only introduced when new coaches became necessary on the various routes. The immediate result was to democratise coach-travelling."

On the other hand, it became a common practice amongst the smart youths of the day to drive the stage-coaches themselves. So we read in a paper of this time:—

> "The education of our youth of fashion is *improving* daily: several of them now drive Stage Coaches to town, and open the door of the Carriage for passengers, while the coachman remains on the box. They *farm* the perquisites from the Coachman on the road, and generally pocket something into the bargain."

Which was, according to the writer, "a fit subject for ridicule on any stage."

The post-chaises were as ubiquitous as ever. The French nobleman, from whose book I have already quoted, entered one so soon as he landed at Dover.

> "The Post," he records, "is not, as on the Continent, an establishment dependent upon the Government; individuals undertake this business; most of the inns keep Post Chaises; they are good Carriages with four wheels, shut close, the same kind as we call in France *diligences de ville*. They hold three persons in the back with ease are narrow, extremely light; well hung, and appear the more easy, because the roads are not paved with stone. The postilions wear a jacket with sleeves, tight boots, and, altogether, their dress is light, and extremely neat; and they are not only civil, but even respectful. On your arrival at the Inn, you are shown into a good room, where a fire is kept in winter, and tea is ready every hour of the day. In five minutes at most, another Chaise is ready for your departure. If we compare these customs with those of Germany, or particularly in the North, where you must often wait whole hours to change horses, in a dirty room, heated by an iron stove, the smell of which is suffocating; or even those of France, where the most part of the post-houses, not being Inns, have no accommodation for travellers, it is evident that the advantage is not in favour of the Continent."

Indeed, England at this time was superior to most European countries so far as her posting-carriages and roads were concerned. Leigh Hunt, in expressing his delight of them, was only following in the wake of Johnson and the others who had always enjoyed their cross-country rides.

"French Travelling, or the First Stage from Dover"

(From a Drawing by Rowlandson, 1785)

"A post-chaise," he says, "involves the idea of travelling which, in company of those we love, is home in motion. The smooth running along the road, the fresh air, the variety of scene, the leafy roads, the bursting prospects, the clatter through a town, the gaping gaze of a village, the hearty appetite, the leisure (your chaise waiting only upon your own movements), even the little contradictions to home-comfort, and the expedients upon which they set us, all put the animal spirits at work, and throw a novelty over the road of life. If anything could grind us young again, it would be the wheels of a post-chaise. The only monotonous sight is the perpetual up-and-down movement of the postilion, who, we wish exceedingly, could take a chair. His occasional retreat to the bar which occupies the place of a box, and his affecting to sit upon it, only remind us of its exquisite want of accommodation. But some have given the bar, lately, a surreptitious squeeze in the middle, and flattened it a little into something obliquely resembling an inconvenient seat."

Prints of these post-chaises are common. Rowlandson, in particular, loved to draw them. Gillray, too, shows the post-chaise in Scotland and Ireland, where apparently things were not quite so easy as in England. The Scottish post-chaise is shown breaking to pieces, and the Irish chaise is little better than a wreck, with the body held together by a piece of rope, with hardly a spoke left to the wheels, and a roof put roughly together of thatched straw. The unfortunate lady inside has put one foot through the panelling and another through the floor, which reminds one that it was of an Irish post-chaise that the famous story of the poor man who had to run with the carriage because the bottom had fallen out was originally told.

It remains to consider a few particular eighteenth-century carriages of other countries.

Early American Shay

(From "Stage Coach and Tavern Days" [A. M. Earle])

English Posting Chariot—Early Nineteenth Century

(From a Photograph)

Mr. Stratton thinks that the Indians of North America had rude litters at an early date. The Incas of Peru certainly possessed magnificently decorated sedans or palanquins, in which they progressed through their kingdom. It was not, however, until the seventeenth century that wheeled carriages appeared in America. Sir Thomas Browne quotes from an English traveller's book, which states that by the middle of this century there were at least twenty thousand coaches in Mexico, and possibly this was true. But into North America carriages filtered but slowly. There had been coaches in Boston so early as 1669, and in Connecticut in 1685. William Penn, writing to Logan in 1700, bids his servants have the coach ready. The calash was also known at that time, but being "clumsy" was less popular than the French cabriolet or gig, which had been brought over by the Huguenots, and rapidly transformed into the well-known *one-horse shay*, which in its turn was supplanted by the more comfortable and certainly more distinctive *buggy*.

Bennet, travelling in America in 1740, saw many carriages in Boston.

"There are several families," he records, "in Boston that keep a coach and a pair of horses, and some few drive with four horses; but for chaises and saddle-horses, considering the bulk of the place, they outdo London. They have some nimble, lively horses for the coach, but not any of that beautiful black breed so common in London.... The country carts and wagons are generally drawn by oxen, from two to six, according to the distance, or the burden they are laden with."

A Boston advertisement of 1743 mentions "a very handsome chariot, fit for town or country, lined with red coffy, handsomely carved and painted, with a whole front glass, the seat-cloth embroidered with silver, and a silk fringe round the seat." This was offered for sale by John Lucas, a local coachbuilder, and had most probably been built by him.

At this time several stage-coaches were running, and the *shay* was being used by even the poorer folk. A Philadelphian advertisement of 1746 speaks of "two very handsome chairs, with very good geers," and at this time, too, the Italian chairs and curricles were also popular. They were generally driven tandem.

Even more distinctive than the shay, however, was the *coachee*, which is described by Isaac Weld in his travels (1795): —

> "The body of it is rather longer than a coach, but of the same shape. In the front it is left quite open down to the bottom, and the driver sits on a bench under the roof of the carriage. There are two seats in it for passengers, who sit in it with their faces to the horses. The roof is supported by small props, which are placed at the corners. On each side of the door, above the panels, it is quite open; and, to guard against bad weather, there are curtains which let down from the roof and fasten to buttons on the outside. The light wagons are in the same construction," he adds, "and are calculated to hold from four to twelve people. The wagon has no doors, but the passengers scramble in the best way they can over the seat of the driver. The wagons are used universally for stage-coaches."

The American stage-waggon is also described by another Englishman, Thomas Twining, who visited the country in 1795.

> "The vehicle," says he, "was a long car with four benches. Three of these in the interior held nine passengers. A tenth passenger was seated by the side of the driver on the front bench. A light roof was supported by eight slender pillars, four on each side. Three large leather curtains suspended to the roof, one at each side and the third behind, were rolled up or lowered at the pleasure of the passengers. There was no place nor space for luggage, each person being expected to stow his things as he could under his seat or legs. The entrance was in front over the driver's bench. Of course, the three passengers on the back seat were obliged to crawl across all the other benches to get to their places. There were no backs to the benches to support and relieve us during a rough and fatiguing journey over a newly and ill-made road."

The body of these public carriages was high, and the back wheels were larger than those in front. A somewhat similar conveyance is still used to-day in some of the northern districts of Australia.

The commonest vehicle in Russia at this time seems to have been the *taranta*, which is described as "a travelling carriage whose body resembles a flat-bottomed punt." The natives apparently considered that it was a very comfortable carriage, and it certainly could hold a great quantity of luggage and wraps, but the foreigners using it did not always express a similar opinion.

"We travelled certainly with speed," says Madame Pfeiffer of the *taranta*, in her *Journey round the World*, "but any one who had not a body of iron, or a well-cushioned spring carriage, would not find this very agreeable, and would certainly prefer to travel slower upon these uneven, bad roads. The post-carriage, for which ten kopecs a station is paid, is nothing more than a very short wooden open car, with four wheels. Instead of a seat some hay is laid in it, and there is just room enough for a small chest, upon which the driver sits. These cars naturally jolt very much. There is nothing to take hold of, and it requires some care to avoid being thrown out. The draught consists of three horses abreast; over the centre one a wooden arch is fixed, on which hang two or three bells, which continually made a most disagreeable noise. In addition to this, imagine the rattling of the carriage, and the shouting of the driver, who is always in great activity urging on the poor animals, and it may be easily understood that, as is often the case, the carriage arrives at the station without the travellers."

Even less "genteel" than the *taranta* was the *kibitka*, "a common posting-waggon," according to Stratton, "consisting of a huge frame of unhewn sticks, fastened firmly upon two axles, the fore part of it having underneath a solid block of hard wood, on which it rests, elevating it so as to allow the wheels to play."

Other Russian carriages were the *teleka*, the *telashka*, and the better-known *droitzschka*, or, as it was known in England, *drosky*—an improvement originally of the sledge by the mere addition of springs and wheels. In Norway the *carriole* was very similar to the original French gig, and like the *char-à-cote* of Switzerland, was long and narrow and peculiarly adapted for mountainous countries. But in nearly all the colder regions, wheel carriages were scarcely used at all, the snow making some kind of sledge far more convenient. Captain King, in his *Journey across Asia*, gives a detailed description of the sledges then in use (1784) in Kamtschatka.

"The body of the sledge," he says, "is about four feet and a half long and a foot wide, made in the form of a crescent, of light, tough wood, strongly bound together with wicker-work; which in those belonging to the better sort of people is elegantly stained of a red and blue colour, and the seat covered with bear-skins, or other furs. It is supported by four legs, about two feet high, which rest on two long flat pieces of wood, extending a foot at each end beyond the body of the sledge. These are turned up before, in the manner of a skate, and shod with the bone of some sea animal. The fore part of the carriage is ornamented with thongs of leather and tassels of coloured cloth; and from the cross-bar, to which the harness is joined, are hung links of iron, or small bells, the jingling of which they conceive to be encouraging to the dogs. They are seldom used to carry more than one person at a time, who sits aside [? astride], resting his feet on the lower part of the sledge, and carrying his provisions and other necessaries, wrapped up in a bundle, behind him. The dogs are usually five in number, yoked two and two, with a leader. The reins not being fastened to the head of the dogs, but to the collar, have little power over them, and are therefore generally hung upon the sledge, whilst the driver depends entirely on their obedience to his voice for the direction of them.... The driver is also provided with a crooked stick, which answers the purpose both of whip and reins; as by striking it into the snow, he is enabled to moderate the speed of the dogs, or even to stop them entirely.... Our party consisted in all of ten sledges. That in which Captain Gore was carried, was made of two lashed together, and abundantly provided with furs and bear-skins; it had ten dogs, yoked four abreast, as had also some of those that were heavy laden with baggage."

In Europe and North America these sledges were also used, and could be highly ornamented. Two of this kind, narrow and low, may be seen at South Kensington. They are mentioned by several travellers. Edward Wright, visiting Amsterdam in 1719, had seen "several coach-bodies drawn upon sledges," and explained that the inhabitants did not use wheels "to avoid shaking the foundations of the houses." Holcroft, too, at the end of the century, journeyed from Hamburg to Paris by way of Holland, and did not hide his surprise at the appearance of these sledges.

"And pray, sir, what are you?" he asks in the Shandean manner. "We never saw so staring or so strange an animal before."

"'Tis a tropical bird, on a mast."

"Can it be? A coach without wheels? Yes: dragged on a sledge by a single horse, and a lady in it."

Holcroft also noticed in Amsterdam what he called "a travelling haberdasher's shop with wheels, rolled through the streets by its master." This appears to have been some sort of light travelling booth. In Paris itself, he records, "there is scarcely a street which is not so narrow as to be extremely dangerous to foot passengers. They are rendered more so at some times by the extreme carelessness, and at others by the brutal insolence, of coachmen. There is no foot pavement; and the only guard against carriages is formed by large stones placed at certain distances, but close to the wall." In Germany, too, he found little to please him, and warns Englishmen against bringing English-built carriages into that country, for of a surety they will be "broken up." England, indeed, about this time, seems to have been by far the most progressive country as regards locomotion.

CHAPTER THE NINTH
INVENTIONS GALORE

"Prime of Life to 'go it!' where's the place like London:
Four-in-hand to-day, tomorrow you may be undone:
Where the Duke and the 'prentice they dress much the same:
You cannot tell the difference, excepting by the name!
Then push along with four-in-hand, while others drive at random,
In buggy, gig or dog-cart, in curricle or tandem."

EGAN, *Life in London.*

IF William Felton's book shows the great improvements that had taken place in English carriage-building during the latter half of the eighteenth century, William Bridges Adams's *English Pleasure Carriages*, published in 1837, sufficiently shows the enormous improvements which had followed upon Obadiah Elliott's invention of the elliptic springs.[48] In the first place you had a whole series of light, <u>perchless carriages</u> being built, and in the second you had the new macadamised

[48] It may be well to add here a note on the simpler springs which were in use at this time. These seem to have been of five distinct varieties — the straight or elbow spring, the elliptic spring, the regular-curved, and the reverse-curved springs, all these being either single or double, and the spiral spring. The straight spring was used in the stage-coaches, in the later phaetons, in the Tilbury, and in most of the two-wheeled carriages. The elliptic spring, invented by Elliott, was "used single in what are called under-spring carriages, where the spring rests on the axle, and is connected with the framework by means of a dumb or imitation spring so as to form a double or complete ellipse. This is technically called an under spring." Its importance, of course, followed on its power of acting as a complete support, no perch being required to hold the two parts of the undercarriage together. Sometimes four of these springs were "hinged together in pairs," and used thus in the larger four-wheeled carriages. When a regular-curved or C spring was used, "a leathern brace was suspended from it to carry the body or weight." The reverse-curved spring was used in the older phaetons, and in the fore springs of the Tilbury, and springs similar to this had been used as body springs in place of suspension brackets or loops, or as upright springs, to the earlier coaches and chariots, under the technical name of S springs — "in which case leather braces were attached to them, and they were supported by a bracket or buttress of iron called the spring stay. The whip spring which succeeded them ... was used in the same way." But in addition to these springs, there were all kinds of combinations, and the whole subject is too complicated for the lay mind to understand. The chief point, however, to notice is the changes in structure which were made possible by the elliptic spring of Elliott's resting on the axle.

roads upon which to run them.

In treating of all these various carriages, it is difficult to know where to begin. A mere catalogue with a few lines of description cannot be very satisfactory, and yet there seems no other method to adopt. Bridges Adams, who was a coach-builder himself and the inventor of several novel carriages, is a good guide, but one could have wished that his book had been illustrated by anything rather than those fearsome diagrams which mean so little to any one but a coachbuilder himself. From the beginning of the century, indeed, illustrations of carriages began to take on that diagrammatic aspect which the trade-papers still maintain; while at the same time the old prints and caricatures began to disappear. It is a pity, but it cannot be helped.

"Though it would be difficult," says Bridges Adams, "to describe every particular variety of carriage now in use, it is comparatively easy to set forth the leading features—the original models, as it were, of each particular class. The distinguishing characteristics are to be found in the form of the bodies and not in the mechanism of the springs or framework. Thus a particular shaped body entitles the carriage to the term Chariot, whether it be constructed with under springs or C springs, or with both, or whether it be with or without a perch. This rule obtains throughout the whole varieties of carriages; and in those bodies which are formed by a combination"—as now began to be the case—"it is customary to call them by a double name—as Cab-Phaeton, Britzschka-Chariot, Britzschka-Phaeton, &c." Accordingly, I shall endeavour in a brief catalogue to point out such changes as were being made in each broad class of vehicle.

The coach was still being made with a perch. It was not hung so high, but in other respects it differed but little from its predecessors. The Salisbury boot, which carried the coachman's seat, and the hammercloth, were still used, but for travelling long distances were removed, a smaller platform being substituted in their place. In the *Driving Coach*, a novelty which now became popular with gentlemen of means, and at a later date came to be commonly known as the *four-in-hand*, the wheels were rather nearer together, and the perch was short and straight. This had the boots which, as we have seen, had been already added to the mail-coaches for the convenience of outside passengers. "The boots and body," says Bridges Adams, "are framed together, and suspended on springs before and behind—the connection with the carriage being by means of curved blocks."

Another variety of the coach was the *barouche*, which, though, I suppose, not technically a coach at all, if one accepts Thrupp's definition—for it was roofless—is generally classed with this kind of vehicle. There had been, I believe, a barouche in England so early as 1767, but it was not popular until a much later date. The barouche was simply a coach-body without its upper portion—an open carriage, that is to say, with high driving seat, and a hood fixed to the back if required—not indeed unlike an opened landau to look at. It was purely a town carriage. Its driving seat, similar to that in a landau, was built to hold both coachman and footman, "the hinder part being unprovided with a standard, which would," says Bridges Adams, "be useless, as when the head is down there is little convenience for the servant's holders, and he would moreover be unpleasantly placed, looking down on the sitters within, and listening to all the conversation," a matter of course which he would have been only too pleased to do. The barouche would hold four or six persons, and in fine weather was considered to be "the most delightful of all carriages." There was, too, a certain amount of state about it, and several noble families continued to drive in them long after most other people had given them up. When Ackermann, the publisher, invented his patent movable axles about 1816, the barouche was one of the carriages to which these axles were fitted. A print of this carriage is shown in the accompanying illustration. A *barouchet*, corresponding to the landaulet, was also built at this time, but was never popular. Bridges Adams speaks of it as a graceless carriage for one horse.

The town chariot, or *coupé*, as it was called in France, and indeed, at a later date in England, was being built lower than before, but otherwise remained unaltered. The high driving seat was still removed to transform the carriage into a post-chaise. Amusing instructions for buying a chariot are given by John Jervis, an old coachman, in the second volume of the *Horse and Carriage Oracle*, 1828. "The form of Carriages," he opines, "is as absurdly at the Mercy of Fashion, as the Cut of a Coat is—however, if the Reader is willing to let the Builder please himself with the form of the Exterior, he will not be quite so polite as to submit the construction of the Interior entirely to the caprice of his Coachmaker." Don't, he advises, have too much stuffing inside: "*The present fashion of Stuffing* is preposterous, it reduces a Large Body to the size of a small One: however," he adds obligingly, "if you like to ride about for the benefit of public inspection, as your friends, my *Lady Look-out*, the *Widow Will-be-seen*—and *Sir Simon Stare*, do, pray, study Geoffrey Gambado on the Art of sitting politely in Carriages, with the most becoming attitudes, &c., and choose wide Door Lights and full Squabbing;—if you wish to go about peaceably and quietly, like *Sir Solomon Snug*, and are contented with seeing without being seen, adopt the contracted Lights, and common Stuffing, which, among others, have this great advantage that when you sit back, you may have the side Window down, and a thorough Air passing through the Carriage, without it blowing directly in upon you: this, to Invalids who easily catch Cold, is very important." The lining of the chariot, he recommends, should be "green, with Lace to correspond, and the Green *silk Sun Shades* of the same Colour," green being pleasant to the eye. Venetian blinds, he says, are very nice in warm weather, and should be painted verdigris green on the inside and on the outside a colour which matches with that of the coach-body. Further instructions follow. You are advised never to permit officious strangers to shut your carriage door—a piece of sound advice which might well be followed to-day when seedy people expect a small tip for having watched you get into a cab—and if your coachman sees any one about to do so, he is to say "loudly and imperatively, '*Don't meddle with the Door!*'"

The chief maker of these chariots was the celebrated Samuel Hobson, "who may be truly said to have improved and remodelled every sort of carriage, which came under his notice, especially as regards the artistic form and construction, both of body and carriage." "Hobson's Chariots," indeed, were in a class by themselves. "He lowered the wheels of coaches and chariots," says Thrupp, "to 3 ft. 3 in. in front and 4 ft. 5 in. behind, and lengthened the carriage part once more to such a true proportion to the whole vehicle as has approved itself as correct to each succeeding generation of Coachbuilders and users of carriages. He lowered the body, too, so that it could be entered by a moderate double step instead of the three-fold ladder previously in use."

Barouche

With Ackerman's Patent Movable Axles

Landaulet

With Patent Roof and Movable Axles

Mr. Jervis's remarks about the coachmaker's being allowed to choose the exterior of his customer's carriage no doubt followed on the practice, mentioned by Bridges Adams, of building particular carriages upon a general chariot basis. Of these hybrids, perhaps the most popular was the *Briska-chariot*. The briska itself (more correctly the *britzschka*) had been introduced into England from Austria about 1818 by Mr. T. G. Adams, though Bridges Adams thinks that it was first brought here at a rather later date by the Earl of Clanwilliam, "who liked it for its lightness; for which reason it probably obtained, amongst coachmen and mechanics, the translated name *Brisker* or *Brisky*." In England it was made in various sizes and with various modifications. A small one for one horse was "a light open carriage, fitted with a leathern top over the front inside seat; which top had a glazed front and sides, or glazed front and Venetian blinds to the sides." Its chief characteristics were a small seat at the back of the main body and a straight bottom line to the body itself—this giving it "a ship-like and fast-going appearance." Ten years after its introduction it was so immensely popular as to threaten every other carriage; nor was this altogether surprising, for in addition to being liked for the sake of its own lightness, it lent itself so well to every variety of purpose. And of these modified briskas, the *briska-chariot* was one of the most favoured. It was in particular demand with those travelling abroad, inasmuch as its great length enabled its passengers to lie at full length. Another variety, the *droitzschka* or *drosky*, was a modification of the Russian vehicle of that name. This was built low, an open perch carriage with a hood, used chiefly by "languid, aged, or nervous persons, and children." The drosky seems to have given the idea to Mr. David Davies for his *pilentum*, which was very similar in appearance. This Mr. Davies is also supposed to have been the inventor of the popular *cab-phaeton*, a one-horse, low-hung carriage suspended on four elliptic springs. On the Continent this carriage became known as a *milord*, once most aristocratic, but by 1850 little better than a hack. It was somewhat similar in appearance to the *victoria*.

The phaeton was still made, but was being superseded by the briska. The main seat of the carriages, as in the old perch-high phaetons, was still over the front axle, but the body was now hung low on elliptic springs. Such a perchless carriage was called by Adams "the very simplest form of wheeled vehicle in ordinary use. It is literally a long box, with an arm-chair in front, and a bench behind." And that is a remarkably good description. Here, too, as with the chariots, there were also various hybrids.

Landaulets were very popular in London, and were made in great quantities by the firm with which Obadiah Elliott himself was connected. A patent roof and Ackermann's movable axles are shown in the accompanying illustration of this carriage.

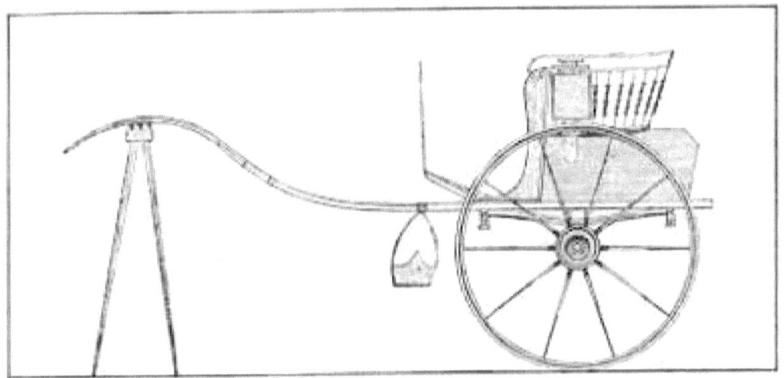

Stanhope

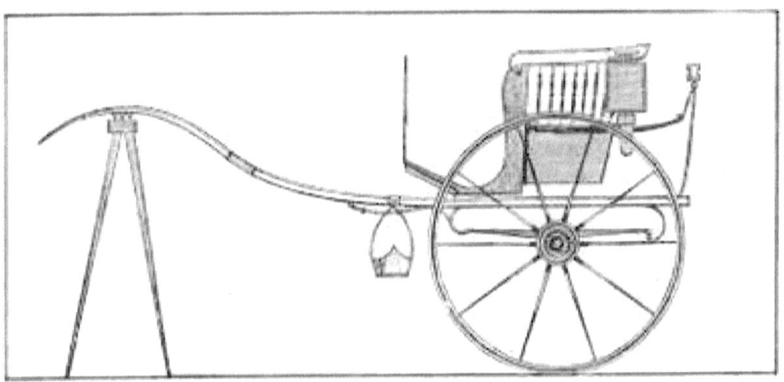

Tilbury

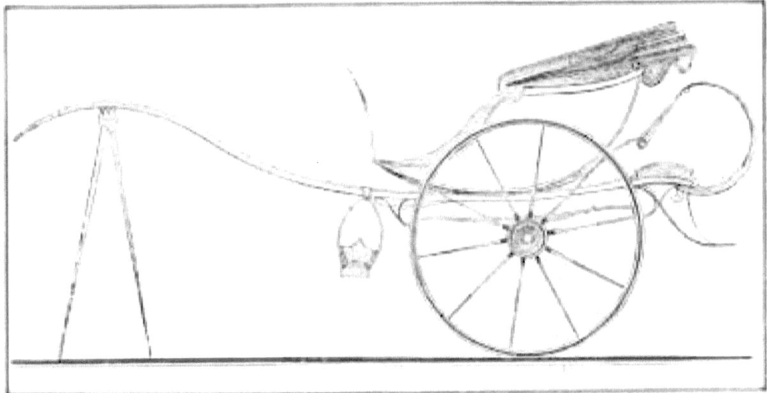

Cabriolet

We come now to the two-wheeled carriages. Of these the most fashionable was still the curricle, though Bridges Adams considered the shape of the body "certainly unsightly." It is interesting to notice in this connection that the mode of attaching the two horses to the curricle was "precisely that of the classic car, only more elegant." It was in a curricle that Charles Dickens rode about so soon as he was able to afford the luxury of a private carriage. The *cabriolet*, somewhat similar to it in form, was simply the old one-horse chaise brought up to date. The body resembled a nautilus shell, thus differing from the popular two-wheeled carriage called a *tilbury*. This had been built first by a carriage-maker of the same name. It was constructed without a boot (or hind seat) and was a very light carriage, with, however, rather too much ironwork and too many springs—seven in all— about it. Italy and Portugal seem to have taken to this particular gig and numerous consignments were sent south by water. Another vehicle, not very different, was the *stanhope*, also built by Tilbury to the order of the Hon. Fitzroy Stanhope, a brother of Lord Petersham. This was much like the old rib chair, but hung from four springs. The only difference, so far as the shape of their bodies goes, between the tilbury and the stanhope is to be found in the fact that in the stanhope it is rather larger and more capacious. The *dennet*, invented by a Mr. Bennett of Finsbury, had a body resembling that of a phaeton. It had three springs, and Bridges Adams, without being certain upon the point, thinks that it took its name from these three springs, which were named after the three Misses Dennet, "whose elegant stage-dancing was so much in vogue about the time the vehicle was first used." The lightest of all these carriages, however, was the common gig, such as that arch-joker, Theodore Hook, was accustomed to drive in, which at this time was "simply an open railed chair, fixed on the shafts, and supported on two side springs, the harder ends of which were connected to the loop irons by leathern braces—to give more freedom to the motion." Small alterations in the gig, such as the addition of a deep boot and Venetian blinds to the lockers (to carry dogs) led to the first *dog-cart*. Here the passengers sat back to back. *Tandem-carts* were very similar, though here the driver's seat was raised. The dog-cart itself gave rise to numerous varieties, such as the *Newport*, the *Malvern*, the *Whitechapel*, the *sliding body*, and the *Norwich* carts.

In America the *buggy*, a light waggon, the *sulky*, the *fantail gig*, the *tub-bodied gig*, the *chariotee*, and the *public sociable* were the chief carriages. The *rockaway*, made first in 1830, was a light waggon with wooden springs on the outside of the body. The *volante*, much used at this time by the Spanish ladies of South America and Cuba, was a hooded gig upon two high wheels. But in America, as in Europe, no entirely new bodies or methods of framing were needed, and such little differences as there were are only of interest to the coachbuilder or the expert.

Before passing, however, to the public conveyances, to which, it would seem, most carriage-builders of an inventive turn were now giving their attention, I may mention one or two particularly quaint or fanciful carriages which do not readily fall into a recognised class.

About this time several people seem to have been at pains to produce a three-wheeled carriage, "apparently designed," says Croal, "to overcome an element of danger in the ordinary two-wheeled gig, in which so much of the business and pleasure of travelling took place." In America, the chief experiments in this direction were made by Dr. Nott, president of Union College at Schenectady, who produced a three-wheeled chariot, in which he drove about.[49] "The body of the vehicle was supported by the near axle on two wheels, while a third wheel in front was in close connection with the shafts, so that it revolved with them as they turned. By this arrangement the body of the carriage could be hung low, supported entirely by the wheels, while the third wheel in front, revolving in a small circle with the shaft, enabled the occupants to make a short and safe turn." What became of this weird vehicle is not known, but its inventor's memory was enshrined in a song, one verse of which runs as follows:—

> "Where, oh where, is the good old Doctor?
> Where, oh where, is the good old Doctor?
> He went up in the Three Wheel Chariot,
> Safe into the Promised Land!"

A six-wheeled carriage was also proposed by Sir Sidney Smith. Here, as in Bridges Adams's various equirotal carriages (never successful and particularly ugly, so far as the pictures of them are concerned), the wheels were all of equal size. Great things were promised of it, but that was all. The question, however, of safety carriages was being very widely considered. Accidents must have been all too frequent. Runaway horses and high gigs between them were constantly bringing the more reckless drivers to an untimely end. In 1825 a good proposal was made for a safety gig, which was to have a contrivance fixed to the shafts so that they should remain in a horizontal position, whether the horse were between them or not. Experiments were also made with some such contrivance as Sir Francis Delavel had first tried with his eighteenth-century phaeton. And then came a time when almost every coachbuilder had some "pet dodge" with which the dangers of travelling were supposed to be reduced to a minimum.

In Ireland, where at a very early date a rough, flat-boarded waggon on two solid wheels had been used for passenger-traffic—in which case the passengers sat on the boards back to back with their legs dangling over the sides—a peculiar vehicle called a *noddy* was now popular. A writer in *Blackwood's Magazine* for 1826 speaks of this carriage.

> "A chaise and pair, miserable in show and substance as both
> really were, was a species of luxurious conveyance to which the
> ambition of the middle class of travellers in Ireland before 1800
> never ventured to aspire. Such as were content with a less dig-
> nified mode of travelling on wheels, the city of Dublin accom-
> modated with a vehicle unparalleled, I believe, in any part of the
> world, and singular in name as well as construction. It was called
> a *Noddy*, drawn by one horse, and carrying two, or if not of over-

---

[49] Which reminds me that at the present day there is a singular three-wheeled cab to be hired in London, if only you know where to look for it. It is the only one of its kind, and rarely, I believe, appears until after nightfall. It is the kind of carriage which is to be avoided by those who have drunk not wisely but too well.

grown dimensions, three passengers. The body of this 'leathern convenience,' which bore some resemblance to an old-fashioned phaeton, 'beetled o'er its base' in front, the better to protect the inmates; and being slung from cross-bars by strong braces instead of springs, nodded formidably at every movement of the horse, hence deriving the appropriate appellation of Noddy. In case of rain blowing in, a curtain of the same material afforded its friendly shelter, wrapping the passengers in total darkness, though, as far as the prospect was concerned, the inconvenience was little; the only visible object when it was withdrawn being the broad back and shoulders of the brawny driver, who rested his legs upon the shaft, and his sitting part on a sort of stool a very little way removed from the knees of the person seated within. Simple, awkward, and uneasy as this contrivance was, it was not disdained even by senators at an earlier period than that of which I write; and a nobleman, some thirty years older than myself, too, of high rank and large estate, assured me that it was his usual conveyance to and from college accompanied by a trusty servant or private tutor."

The ordinary *jaunting car* and the larger *bian*—the invention of Bianconi, a rich tradesman in Dublin, though for many years an itinerant dealer—hardly differed in points of construction from English carriages, though the passengers sat back to back on a seat that ran parallel to the shafts.

In Wales the *market cart* was even more primitive than the noddy of Ireland. This was a low, two-wheeled, springless box of an affair, in which you sat as best you could on the boards. There was no covering at all. A rail at the back, extending some way along the sides, helped to prevent you from falling out behind, if the horse gave a sudden lurch forward.

Whilst European carriages were thus taking on a soberer aspect, Eastern coaches were maintaining all their old magnificence. The Maharajah of Mysore, to take one instance, travelled in a truly marvellous elephant carriage in the early years of the nineteenth century.

"Its interior was a double sofa for six persons, covered with dark green velvet and gold, surmounted by an awning of cloth of gold, in the shape of two small scalloped domes, meeting over the centre, and surrounded by a richly ornamented verandah, supported by light, elegant, fluted gilt pillars. The whole was capable of containing sixty persons, and was about twenty-two feet in height. It moved on four wheels, the hinder ones eight feet in diameter, with a breadth of twelve feet between them. It was drawn by six immense elephants, an exact match in size, with a driver on each, harnessed to the carriage by traces, as in England, and their huge heads covered with a sort of cap made of richly embroidered cloth. The pace at which the elephants moved was a slow trot, of about seven miles an hour—they were very steady, and the springs of the coach particularly easy. The shape of the body was that of an extremely elegant flat scallop-shell, painted dark green and gold. This magnificent carriage was the production of native workmen, assisted by a half-caste Frenchman."

Even this vehicle, however, was eclipsed by the state carriage of a ruling Burmese chief, captured by the British in 1824. "This carriage presented one entire blaze of gold, silver, and precious stones; the last-named amounting to many thousands, including diamonds, rubies, blue and white sapphires, emeralds, amethysts, garnets, topazes, crystals, and the curious and rare stones known as cat's eyes. The carriage stood nearly thirty feet in height," and was drawn by elephants. "In form and construction," says Croal, "in its elaborate and superior carving, and its grand and imposing effect, this coach takes rank as one of the most splendid equipages in existence."

Many changes, meanwhile, were taking place in the public carriages.

Of the mail-coaches I need say nothing at all. Numerous books exist which retell all those romances of the road which even in these days of motor-cars cannot be altogether forgotten. The Golden Age of coaching was at hand, and no print-shop is complete without some score or more of carefully coloured engravings of one or other of "the Mails." They bore particular names—there were Flying Machines and Telegraphs and the like—and they were larger than in the days when Palmer had inaugurated the system, but that was all.[50]

Coming to such public vehicles, however, as were in general confined to the metropolis, we find many changes.

The old hackney-coaches still plied for hire. They had their particular stands, and the fares were subject to strict, though sometimes exceedingly quaint, regulations. The first section of the new *Orders* issued in 1821 may be quoted as bearing

[50] A good description is given of the appearance of these coaches by Baron d'Haussez, an exiled Frenchman, in 1833.

"The appointments of an English coach are no less elegant than its form. A portly, good-looking coachman seated on a very high coach-box, well dressed, wearing white gloves, a nosegay in his button-hole, and his chin enveloped in an enormous cravat, drives four horses perfectly matched and harnessed, and as carefully groomed as when they excited admiration in the carriages of Grosvenor and Berkeley Squares. Such is the manner in which English horses are managed, such also is their docility, the effect either of temperament or training, that you do not remark the least restiveness in them. Four-horse coaches are to be seen rapidly traversing the most populous streets of London, without occasioning the least accident, without being at all inconvenienced in the midst of the numerous carriages which hardly leave the necessary space to pass. The swearing of ostlers is never heard at the relays any more than the neighing of horses; nor are you interrupted on the road by the voice of the coachman or the sound of his whip, which differs only from a cabriolet whip in the length of the thong, and serves more as a sort of appendage than a means of correction in the hand which carries it."

upon the structure of the hackneys.

"It is ordered, constituted, and ordained, that, from and after the four-and-twentieth Day of *June* next ensuing the Day of the Date of these Presents, the Perch of every Coach shall be Ten Feet long at the least; and such Coach [shall] have cross Leather Braces before, and not braced down, but shall hang upon a Level, and not higher behind than before, and to be decent, clean, strong, and warm, with Glass Windows on each Side, or Shutters with Glasses of Nine Inches in Length, and Six Inches in Breadth in each Shutter; and large enough to carry Four Persons conveniently; and the Horses to every such Coach shall be able and sufficient for the Business when such Coach and Horses come from Home, to Ply; on a Penalty not exceeding Ten Shillings, at the Discretion of the said Commissioners, to be paid by the Owner of the License, if the same be not rented out, and in Case the same shall be rented out, then upon a Renter thereof."

Leigh Hunt could find little good to say of them. Says he, quoting from a supposititious poetess:—

"Thou inconvenience! thou hungry crop
For all corn! thou small creeper to and fro
Who while thou goest ever seem'st to stop,
And fiddle-faddle standest while you go;
I'the morning, freighted with a weight of woe,
Unto some Lazar-house thou journiest,
And in the evening tak'st a double row
Of dowdies, for some dance or party drest,
Besides the goods meanwhile thou movest east and west.

"By thy ungallant bearing and sad mien,
An inch appears the utmost thou couldst budge;
Yet at the slightest nod, or hint, or sign,
Round to the curb-stone patient dost thou trudge;
School'd in a beckon, learned in a nudge;
A dull-eyed Argus watching for a fare;
Quiet and plodding, thou doest bear no grudge
To whisking Tilburies, or Phaetons rare,
Curricles, or Mail-coaches, swift beyond compare."

Dickens was familiar with these hackneys, and in one of the *Sketches by Boz* draws a picture of them.

"Take a regular, ponderous, rickety, London hackney-coach, of the old school, and let any man have the boldness to assert, if he can, that he ever beheld any object on the face of the earth which at all resembles it unless, indeed, it were another hackney-coach of the same date. We have recently observed on certain stands, and we say it with deep regret, rather dapper green chariots, and coaches of polished yellow, with four wheels of the same colour as the coach, whereas it is perfectly notorious to every one who has studied the subject, that every wheel ought to be of a different colour, and a different size. These are innovations, and, like other miscalled improvements, awful signs of the restlessness of the public mind, and the little respect paid to our time-honoured institutions. Why should hackney-coaches be clean? Our ancestors found them dirty, and left them so. Why should we, with a feverish wish to 'keep moving,' desire to roll along at the rate of six miles an hour, while they were content to rumble over the stones at four? These are solemn considerations. Hackney-coaches are part and parcel of the law of the land; they were settled by the Legislature; plated and numbered by the wisdom of Parliament.

"Then why have they been swamped by cabs and omnibuses? Or why should people be allowed to ride quickly for eightpence a mile, after Parliament had come to the solemn decision that they should pay a shilling a mile for riding slowly? We pause for a reply—and, having no chance of getting one, begin a fresh paragraph....

"There is a hackney-coach stand under the very window at which we are writing; there is only one coach on it now, but it is a fair specimen of the class of vehicles to which we have alluded—a great, lumbering, square concern, of a dingy yellow colour (like a bilious brunette), with very small glasses, but very huge frames; the panels are ornamented with a faded coat of arms, in shape something like a dissected bat, the axle-tree is red, and the majority of the wheels are green. The box is partially covered by an old great-coat, with a multiplicity of capes, and some extraordinary-looking clothes; and the straw, with which the canvas cushion is stuffed, is sticking up in several places, as if in rivalry of the hay, which is peeping through the chinks in the boot. The horses with drooping heads, and each with a mane and tail as scanty and straggling as those of a worn-out rocking-horse, are standing patiently on some damp straw, occasionally wincing, and rattling the harness; and now and then, one of them lifts his mouth to the ear of his companion, as if he were saying in a whisper, that he should like to assassinate the coachman. The coachman himself is in the watering-house; and the waterman, with his hands forced into his pockets as far as they can possibly go, is dancing the 'double shuffle,' in front of the pump, to keep his feet warm....

"Talk of cabs! Cabs are all very well in cases of expedition, when it's a matter of neck or nothing, life or death, your temporary home or your long one. But, besides a cab's lacking that gravity of deportment which so peculiarly distinguishes a hackney-coach, let it never be forgotten that a cab is a thing of yesterday, and that he never was anything better. A hackney-cab had always been a hackney-cab, from his first entry into life; whereas a hackney-coach is a remnant of past gentility, a victim to fashion, a hanger-on of an old English family, wearing their arms, and in days of yore, escorted by men wearing their livery, stripped of his finery, and thrown upon the world, like a once-smart footman when he is no longer sufficiently juvenile for his office, progressing lower and lower in the scale of four-wheeled degradation, until at last it comes to—a *stand!*"

These new cabs, indeed, were, as Dickens says, a thing of yesterday, but they had had ancestors. Their immediate forefathers came from Paris, where they had been known for some time under the name of *cabriolets de place*. Light two-wheeled carriages, these were, which had been evolved quite naturally from the original French gig of the seventeenth century. The popularity of these cabriolets in Paris naturally led certain enterprising people in London to attempt their importation, but there was a difficulty to be surmounted. The proprietors of the hackney-coaches had secured a monopoly for carrying people in the streets of London. In 1805, however, licences were obtained for nine cabriolets, which thereupon started to run. In these two passengers could be carried, and the driver sat side by side with his fares.

They were not a great success. In the first place they were not allowed except in certain areas, and in the second passengers did not apparently appreciate the close proximity of the driver. A number of years passed before they either increased in numbers or caught the public fancy. But in 1823, the Mr. Davies who had designed the cab-phaeton built twelve new cabriolets, which were put on to the streets for hire at the end of April.

"'Cabriolets,' runs a newspaper account, 'were, in honour of His Majesty's birthday, introduced to the public this [April 23rd] morning. They are built to hold two persons inside besides the driver (who is partitioned off from his company), and are furnished with a book of fares for the use of the public, to prevent the possibility of imposition. These books will be found in a pocket hung inside the head of the cabriolet. The fares are one-third less than hackney-coaches.'"

These new cabs, painted yellow, had one novel feature which must have astonished the inhabitants, for the driver's seat was a rather comical affair at the side—entirely *outside* the hood. In this way privacy was ensured, particularly if the curtains in front of the hood were drawn together. "The hood," says Mr. Moore,[51] "strongly resembled a coffin standing on end, and earned for the vehicle the nickname of 'coffin-cab.'" Cruikshank's picture of one of these, to illustrate a Sketch by Boz, shows the curious shape of the hood very well. In a short while these cabriolets became popular—there were over one hundred and fifty of them in 1830—particularly with the younger generation. A verse of a then popular song mentions them:—

[51] *Omnibuses and Cabs.*

"In days of old when folks got tired,
A hackney-coach or a chariot was hired;
But now along the streets they roll ye
In a *shay* with a cover called a *cabrioly*,"

which hints at a slightly incorrect pronunciation! But in a short while the cockney found it easier to say *cab*, did so, and has done so ever since.

The Coffin-Cab

(From a Drawing by Cruikshank)

Dickens describes these cabs in his essay on the London streets:—

"Cabs, with trunks and band-boxes between the drivers' legs and outside the apron, rattle briskly up and down the streets on their way to the coach-offices or steam-packet wharfs; and the cab-drivers and hackney-coachmen who are on the stand polish up the ornamental part of their dingy vehicles—the former wondering how people can prefer 'them wild beast cariwans of homnibuses, to a riglar cab with a fast trotter,' and the latter admiring how people can trust their necks into one of 'them crazy cabs, when they can have a 'spectable 'ackney cotche with a pair of 'orses as von't run away with no vun'; a consolation unquestionably founded on fact, seeing that a hackney-coach horse never was known to run at all, 'except,' as the smart cabman in front of the rank observes, 'except one, and he run back'ards.'"

London Cab of 1823, with Curtains drawn

(From "Omnibuses and Cabs")

There is another sketch of Dickens which merits quotation here. The two-wheeled cabs were, of course, soon superseded by others of more modern appearance, and Dickens speaks of the last of the cab-drivers and his particular cab, with a few instructions upon riding in it.

This cabriolet "was gorgeously painted—a bright red; and wherever we went, City or West End, Paddington or Holloway, North, East, West, or South, there was the red cab, bumping up against the posts at the street corners, and turning in and out, among hackney-coaches, and drays, and carts, and waggons, and omnibuses, and contriving by some strange means or other, to get out of places which no other vehicle but the red cab could ever by any possibility have contrived to get into at all. Our fondness for that red cab was unbounded. How we should have liked to have seen it in the circle at Astley's!...

"Some people object to the exertion of getting into cabs, and others object to the difficulty of getting out of them; we think both these are objections which take their rise in perverse and ill-conditioned minds. The getting into a cab is a very pretty and graceful process, which, when well performed, is essentially melodramatic. First, there is the expressive pantomime of every one of the eighteen cabmen on the stand, the moment you raise your eyes from the ground. Then there is your own pantomime in reply—quite a little ballet. Four cabs immediately leave the stand, for your especial accommodation; and the evolutions of the animals who draw them are beautiful in the extreme, as they grate the wheels of the cabs against the curb-stones, and sport playfully in the kennel. You single out a particular cab, and dart swiftly towards it. One bound, and you are on the first step; turn your body lightly round to the right, and you are on the second; bend gracefully beneath the reins, working round to the left at the same time, and you are in the cab. There is no difficulty in finding a seat: the apron knocks you comfortably into it at once, and off you go.

"The getting out of a cab is, perhaps, rather more complicated in its theory, and a shade more difficult in its execution. We have studied the subject a good deal, and we think the best way is to throw yourself out, and trust to chance for alighting on your feet. If you make the driver alight first, and then throw yourself upon him, you will find that he breaks your fall materially. In the event of your contemplating an offer of eightpence, on no account make the tender, or show your money, until you are safely on the pavement. It is very bad policy attempting to save the fourpence. You are very much in the power of a cabman, and he considers it a kind of fee not to do you any wilful damage. Any instruction, however, in the art of getting out of a cab is wholly unnecessary if you are going any distance, because the probability is that you will be shot lightly out before you have completed the third mile.

"We are not aware of any instance on record in which a cab-horse has performed three consecutive miles without going down once. What of that? It is all excitement. And in these days of derangement of the nervous system and universal lassitude, people are content to pay handsomely for excitement; where can it be procured at a cheaper rate?"

Thomas Hood also mentions both hackney-coaches and cabs in one of his comic poems, *Conveyancing*.

"O, London is the place for all
In love with loco-motion!
Still to and fro the people go
Like billows of the ocean;
Machine or man, or caravan,
Can all be had for paying,
When great estates, or heavy weights,
Or bodies want conveying.
"There's always hacks about in packs,
Wherein you may be shaken,
And Jarvis is not always *drunk*,
Tho' always *overtaken*;
In racing tricks he'll never mix,
His nags are in their last days,
And slow to go, altho' they show
As if they had their *fast days*!

"Then if you like a single horse,
This age is quite a *cab-age*,
A car not quite so small and light
As those of our Queen *Mab* age;
The horses have been *broken well*,
All danger is rescinded,
For some have *broken both their knees*,
And some are *broken-winded*."

While these cabs were still running, several experiments were being made with patent carriages. One of these, placed on the streets for a short while, was the invention of Mr. William Boulnois. "It was a two-wheeled closed vehicle," says Mr. Moore, "constructed to carry two passengers sitting face to face. The driver sat on a small and particularly unsafe seat on the top of it, and the door was at the back. It was, in fact, so much like the front of an omnibus that it was well known as the *omnibus slice*. Its popular name was the *back-door cab*. Superior people called it a *minibus*. This cab was quickly followed by a very similar, although larger, vehicle invented by Mr. Harvey. It was called a *duobus*." These two cabs cannot have been very comfortable; the shafts were too short, and the knowledge that a possibly heavy coachman was sitting just above your head seems to have militated against their success.

Another cab, not wholly successful in itself, led the way to the widely popular *hansom*. This was a carriage invented in 1834 by Mr. Aloysius Hansom, the architect of the Birmingham Town Hall. Here the body was "almost square and hung in the centre of a square frame." The driver, as before, sat on the roof, but had a small seat fixed there for his convenience. The doors were in front, on either side of the driver's seat. And the wheels were of a prodigious height—being seven feet six inches. Mr. Hansom, who had obviously seen one of Francis Moore's patent carriages of 1790,[52] himself drove this carriage from Hinckley in Leicester-

[52] See note on p. 115.

shire to London, and found financial support from Mr. Boulnois. Further experiments were made—in one model you had to enter the carriage actually *through* the wheels, the door being in this case at the sides—and it was found that the wheels could be made considerably smaller without danger or inconvenience. Whereupon a company was formed to purchase the invention for a sum of ten thousand pounds. Hansom, however, obtained no more than three hundred, the balance being used to perfect the far from satisfactory cabs which had been placed on the streets. Such improvements as were carried out were the work of Mr. John Chapman,[53] then secretary to the Safety Cabriolet and Two-Wheel Carriage Company, who produced a much safer vehicle, afterwards purchased by Hansom's company. This new cab was placed on the streets in 1836, and proved such a success that it was imitated by numerous other companies. Legal proceedings were instituted, but proved both expensive and not particularly successful, and the "pirate" cabs were allowed to flourish as best they could.

Then, in 1836, was made the first of those four-wheeled cabs,[54] which were not really cabs at all, but which will never be known by any other name. The first of these was built by the ingenious Mr. Davies. It bore superficial resemblance to the chariot. Two passengers could ride inside, and a third on the box at the coachman's side. At this date the old two-wheeled cabs were "a source of acknowledged disgrace, of many alarming accidents, and of lamentable loss of life," and a company was formed to provide "a cheap, expeditious, safe, and commodious mode of conveyance in lieu of the present disgraceful and ill-conducted cabriolets." Two years later Lord Brougham was so pleased with the appearance of these new cabs that he ordered one for his own use. So was the first *brougham* constructed—the earliest private four-wheeled closed carriage to be drawn by a single horse.

> "The original brougham," says Sir Walter Gilbey,[55] "differed in many particulars of design, proportion, construction, and finish from the modern carriage. The body ... was several inches wider in front than at the back, and though both larger and heavier, was neither so comfortable nor so convenient.... [It] was held together by heavy, flat iron plates throughout, and the front boot was connected with the front pillars by strong outside iron stays, fixed with bolts. The wheels were at once smaller in diameter and much heavier. [The carriage] carried a large guard or 'opera board' at the back of the body to protect the occupants from risk of injury in a crush, when the pole of a carriage behind might otherwise break through the back panel—an accident now occasionally seen

---

[53] According to Mr. Moore, whose account of this matter seems perfectly clear, the actual vehicle which proved so popular when plying the streets contained very much more of Chapman's work than of Hansom's, and, indeed, if full justice had been done, these light carriages should have come down to posterity as *chapmans* and not *hansoms* at all. On the other hand it is quite possible, that but for Hansom's work, Chapman would never have given such careful attention to this class of vehicle.

[54] It seems, however, that so long as ten years before *one-horse cars* of this form had been plying for hire in Birmingham and Liverpool.

[55] *Modern Carriages.* Sir Walter Gilbey, Bart. London, 1905.

in our crowded streets. Like all other carriages of the time there was a sword case in the back panel for weapons. It was painted olive green, a very fashionable colour at that period."

Roch's Patent Dioropha, 1851

Brougham in 1859

Another hansom, the *tribus*, may be noticed here, though it was not invented until 1844. In this carriage the driver's seat was at the back on a level with the roof, and the door to his left at the back—the reason of this being that the driver could open or close it without leaving his seat. Another peculiarity was the presence of five windows, two in front, one at either side, and a fifth at the back underneath the driver's seat. The *tribus* was the invention of Mr. Harvey, who also built a *curricle tribus*, for two horses, but neither was successful. The *quartobus* (1844) of Mr. Okey, a four-wheeled vehicle to hold four inside passengers, was likewise withdrawn after a short trial.

A word may here be said of the *omnibus*, which had been introduced in 1819 into Paris, though not under that name, by M. Jacques Laffitte. It was a modern outcome of the old *gondola*. Nine years later the modern name was given to it by M. Baudry, a retired military officer. Laffitte had rivals, and ultimately determined to triumph over them by building a superior vehicle. At this time one of the most celebrated coachbuilders in Paris was an Englishman, once in the Navy, named George Shillibeer. To him came Laffitte, and Shillibeer, whilst at work on the new conveyance, conceived the idea of starting a similar one in London. Accordingly he shipped one over and ran it from Paddington to the Bank. This first omnibus of his was a long, much be-windowed, four-wheeled carriage with a door at the back, and not unlike a private omnibus of to-day. A top-hatted coachman sat on a high seat in front and drove three horses abreast. This was in 1819, and from that time, in spite of the usual opposition, these new and rather unsightly vehicles increased in numbers until there were forty or fifty routes in London alone upon which they were to be hourly seen. A song sung with great success at a time when Shillibeer was extending his operations, particularly in the direction of Greenwich, whither it was proposed to run one of the new railroads, may be quoted:—

> "By a Joint-Stock Company taken in hand,
> A railroad from London to Greenwich is plann'd,
> But they're sure to be beat, 'tis most certainly clear,
> Their rival has got the start—George Shillibeer.

> "I will not for certainty vouch for the fact,
> But believe that he means to run over the Act
> Which Parliament pass'd at the end of last year,
> Now made null and void by the new Shillibeer.

> "His elegant *omnis*, which now throng the road,
> Up and down every hour most constantly load;
> Across all the three bridges how gaily appear
> The *Original Omnibus*—George Shillibeer.

> "These pleasure and comfort with safety combine,
> They will neither blow up nor explode like a mine;
> Those who ride on the railroad might half die with fear—
> You can come to no harm in the new Shillibeer.

> "How exceedingly elegant fitted, inside,
> With mahogany polished—soft cushions—beside
> Bright brass ventilators at each end appear,
> The latest improvements in the new Shillibeer.

"Here no draughts of air cause a rick in the neck,
Or huge bursting boilers blow all to a wreck,
But as safe as at home you from all danger steer
While you travel abroad in the gay Shillibeer.

"Then of the exterior I safely may say
There never was yet any carriage more gay,
While the round-tire wheels make it plainly appear
That there's none run so light as the smart Shillibeer.

"His conductors are famous for being polite,
Obliging and civil, they always act right,
For if just complaint only comes to his ear,
They are not long conductors for George Shillibeer.

"It was meant that they all should wear dresses alike,
But bad luck has prompted the tailors to strike.
When they go to their work, his men will appear
A la Française, Conducteur à Mons. Shillibeer.

"Unlike the conductors by tailors opprest,
His horses have all in new harness been drest:
The cattle are good, the men's orders are clear,
Not to gallop or race—so says Shillibeer.

"That the beauties of Greenwich and Deptford may ride
In his elegant omni is the height of his pride—
So the plan for a railroad must soon disappear
While the public approve of the new Shillibeer."

CHAPTER THE TENTH
MODERN CARRIAGES

"Soon shall thy arm, unconquered steam, afar
Drag the slow barge, or urge the rapid car;
Or on wide waving wings expanded bear
The flying chariot through the realms of air."

Erasmus Darwin.

T HE year of Queen Victoria's Coronation saw the successful opening of the London and Birmingham Railway, and from that time all but a few obstinate folk recognised the fact that the horse as a necessary adjunct to cross-country travelling was doomed. For some time, indeed, certain ingenious gentlemen had been carrying out a number of experiments with self-propelled carriages. Fifteen years before, several inventors had produced cumbrous machines which, without requiring rails, were able to progress along the roads at speeds which compared favourably with those attained by the ordinary coaches. Sir Goldsmith Gurney—to mention, perhaps, the most prominent of these men—had patented a steam-carriage in 1827 which, in spite of attacks from an irate populace who feared machinery as they feared the devil, was quite successful enough to lead the enterprising Mr. Hanning to ask for, and obtain, permission to run similar machines on many of the principal roads of England. Indeed, for a short while, there seems to have been a regular service of these primitive automobiles. Many people, it is true, fought shy of Gurney's boilers, which in spite of the fact that they had been "constructed upon philosophical principles" occasionally exploded. It was after such an explosion at Glasgow that Tom Hood seized the opportunity to write the following lines:—

"Instead of *journeys*, people now
May go upon a *Gurney*,
With steam to do the horses' work
By power of attorney:
"Tho' with a load it may explode
And you may all be undone;
And find you're going up to Heaven
Instead of up to London."

Similarly, many people declared their intention of never patronising the railroads.

Steam, however, had come to stay, and the days of coaching were already numbered.

The net result of the new state of things, so far as private carriages were concerned, seems to have been that the coachbuilders set themselves to perfect the urban vehicles, which became lighter, soberer, and more various. New and less conventional "models" were constantly being exhibited, while for those who could not afford more than a single carriage adaptable bodies were devised. So you might order a vehicle which with small trouble could be entirely changed in appearance. The older dignity, moreover, was giving place to a new smartness. "Carriage people" still formed a class, but families which before had been satisfied to use such public conveyances as there had been, now drove forth in one or other of the cheaper private carriages which were being constructed particularly for their convenience. The dog-cart, for instance, had become common and was undergoing various metamorphoses, and the brougham was rapidly becoming the most popular of all town vehicles. In country lanes, too, appeared the waggonette and its kind. Nothing, indeed, was quite so light as the American buggy with its shallow dish of a body and its extraordinarily thin wheels, but there was no longer that heaviness of line which gives to the older carriages what is to modern eyes such an uncomfortable appearance.

So in 1860 a London coachbuilder could write to the American author of *The World on Wheels*:—

> "Ten years have completed a total revolution in the carriage trade in England. Not only have the Court and the nobility adopted economical habits, and insisted on cheap carriages, but they carry no luggage, as was formerly the case when carriages had to sustain great weight, both of passengers and luggage. The cumbrous Court carriages of former times are being gradually abolished, and instead of the rich linings, laces, fringes, and elaborate heraldry usual to the carriages of the nobility, light vehicles, furnished only with a crest, are used. The changes in construction, and consequent depreciation of stock, were a heavy blow to the master coachbuilders; many of the large houses must have lost, in this manner, from ten to twenty thousand pounds. The trade, having recovered from this blow, is in a more healthy state. The favourite carriages in England at this time were waggonettes, sociables, Stanhope and mail phaetons, basket phaetons and landaus."

> I may speak first of the state or "dress" carriages. "These vehicles," says Thrupp, "had long passed the period in which beautiful carving and elegant painting had been used to disguise, as far as possible, the clumsy state carriages of the eighteenth century. Ever since the building of the Irish Lord Chancellor's state coach by Hatchett or Baxter in 1790, coachbuilders had endeavoured to produce a graceful outline of body, of a fair size no larger than was necessary; the C-springs had been made of a perfect curve,

the perch followed the sweeps of the body, the carving was reduced to a moderate amount, the ornamental painting was confined to the stripes upon the wheels, and the heraldic bearings of the owners of the carriages were beautifully emblazoned on the panels. For further ornament they relied on plated work in brass or silver round the body and on loops and wheel hoops. In every capital of Europe such carriages had superseded the old style, and London and Paris had supplied other countries with most of these state carriages."

At the Queen's Coronation in 1838, Londoners had a good opportunity of seeing these dress carriages, a number of which early in the day were lined up in Birdcage Walk. Most of these belonged to the various ambassadors. The one which excited the widest admiration belonged to Marshal Soult—a French-built carriage, originally built for one of the Royal family. Thrupp describes it. "The body had four upper quarter glasses, with a very elegant deep and pierced cornice of silver round the roof; there were four lamps with large coronets on the tops, and the coach bore a coronet on the roof also. The colouring of the painting was a lovely blue, such as was then called Adelaide blue;[56] this had been varnished with white spirit varnish, and seemed almost transparent in lustre. The whole coach was ornamented with silver and was finished in great taste." Other particulars of these carriages are to be found in the contemporary newspaper reports. We are told of the enormous prices paid. Count Strogonoff purchased for £1600 the carriage which had originally been built at a cost of £3000 by the Duke of Devonshire for his state visit to St. Petersburg. Another ambassador, finding that it was too late to buy a carriage, hired one from one of the Sheriffs at a cost of £250 for the occasion, which strikes one as an excessive price even for Coronation Day.

Modern state carriages retain all their former magnificence with little if any of the old cumbersome and unnecessary ornament. One of the finest examples of this kind of carriage is the state landau built for King Edward and used by him in the Coronation procession.

"This magnificent example of the coachbuilder's art," says Sir Walter Gilbey, "is over eighteen feet long. The body is hung upon

[56] Bridges Adams has an amusing passage on the question of colour. He had his own ideas upon the best colours to use on a carriage body. "For bright sunny days," he thinks, "the straw or sulphur yellow is very brilliant and beautiful; but for the autumnal haze, the rich deep orange hue conveys the most agreeable sensations. The greens used are of innumerable tints, commencing with the yellowish olive, and gradually darkening till they are barely distinguishable from black. Neither apple green, grass green, sea green, nor any green of a bluish tint, can be used in carriage painting with good effect as a ground colour; but in some species of light carriages a pleasing effect may be produced for summer by the imitation of the variegated grasses." Quite a poetical idea! "Blues," he continues, "were formerly principally used as a ground colour for bodies, to contrast with a red carriage and framework. Of late very dark blues have been used as a general ground colour, and when new they are very rich, being a glazed or partially transparent colour; but they very soon become worn and faded, the least speck of dust disfiguring them. Blue is also a cold colour, and while it is unfitted for summer by reason of its easy soiling, it is unpleasant in winter, owing to its want of warmth."

C-springs by strong braces covered with ornamentally stitched morocco; each brace is joined with a massive gilt buckle with oak leaf and crown device. Between the hind springs is a rumble for two footmen; there is no driving seat, as the carriage is intended to be drawn only by horses ridden postilion. The panels are painted in purple lake considerably brighter than is usual in order to secure greater effect; marking the contours of the body and the outlines of the rumble are mouldings in wood carved and gilt, the design being one of overlapping oak leaves.

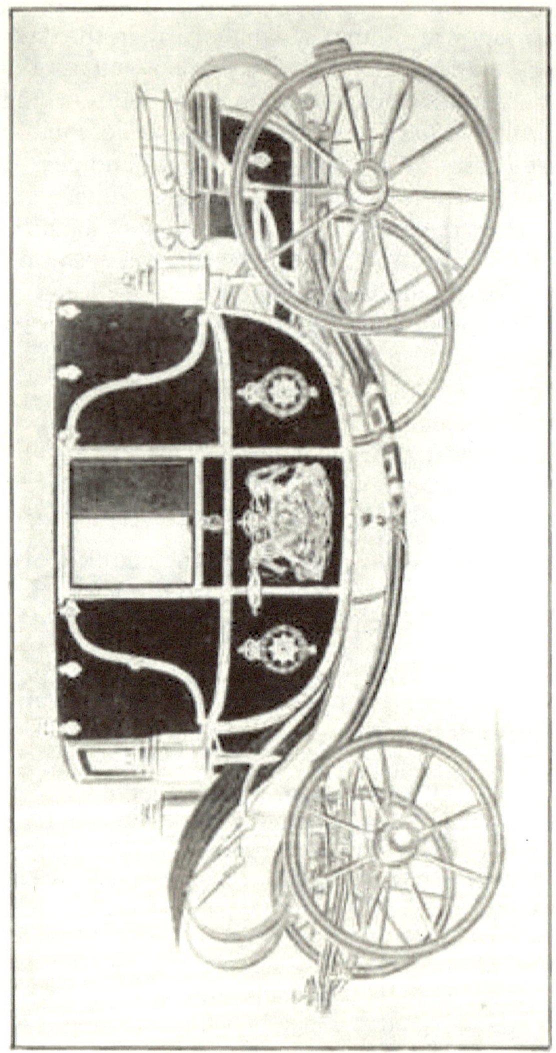

Edward VII's Coronation Landau

(From Sir Walter Gilbey's "Modern Carriages")

"The door panels, back and front panels, bear the Royal Arms with crown, supporters, mantle, motto, helmet, and garter. On the lower quarter panel is the collar of the Order of the Garter, encircling its star and surmounted by the Tudor crown. Springing in a slow, graceful curve from the underpart of the body over the forecarriage is a 'splasher' of crimson patent leather. Ornamental brass lamps are carried in brackets at each of the four corners of the body.

"As regards the interior of this beautiful carriage, it is upholstered in crimson satin and laces which were woven in Spitalfields; the hood is lined with silk, as better adapted than satin for folding. The rumble is covered with crimson leather. It is to be observed that with the exception of the pine and mahogany used for the panels, English-grown wood and English-made materials only have been used throughout.

"While less ornate than the wonderful 'gold coach' designed by Sir William Chambers and Cipriani in 1761, the new state landau, in its build, proportions, and adornment, is probably the most graceful and regal vehicle ever built."

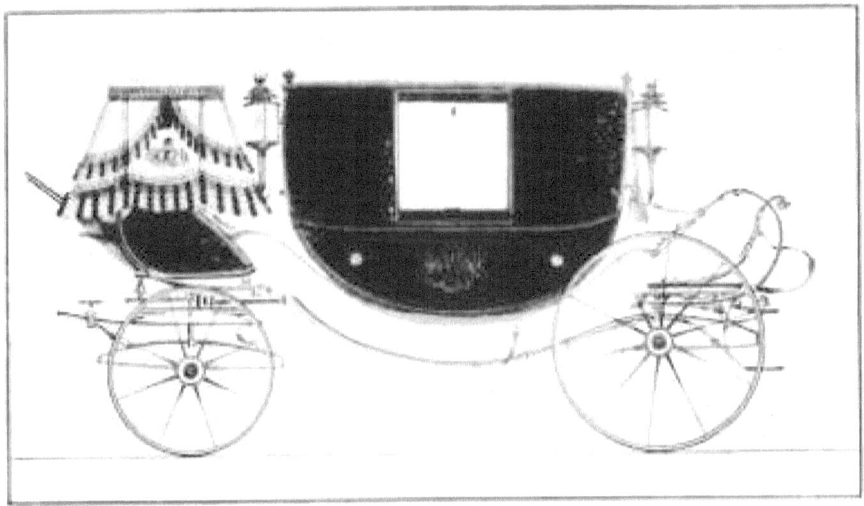

Dress Coach, 1860

Other English state carriages hardly less successfully designed have been made for the Lord Mayor of London (1887), for Sir Marcus Samuel, when holding that position in 1902-3, for the Sheriffs, and for various Indian Princes.

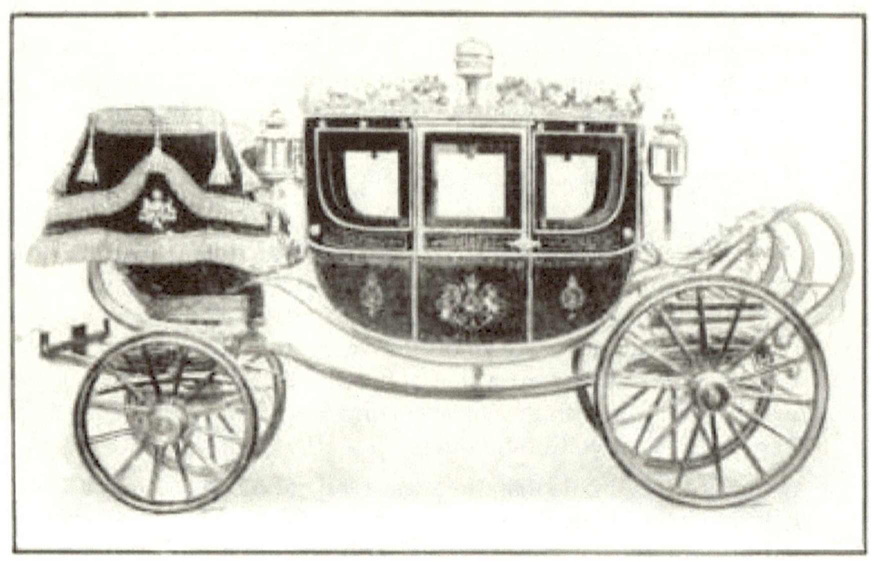

English State Carriage, 1911

(From a Photograph)

Coming to less pretentious vehicles, we may briefly consider in the first place the coach proper. At the time of Queen Victoria's Coronation, coaches of the old pattern were, of course, still being constructed. There is in possession of Messrs. Holland and Holland a mail-coach built by Waude, one of the best-known coach-builders of that time, which is typical of the period. This, says Mr. Charles Harper,

"is substantially and in general lines as built in 1830. The wheels have been renewed, the hind boot has a door at the back, and the interior has been relined; but otherwise it is the coach that ran when William IV was King. It is a characteristic Waude coach, low-hung, and built with straight sides, instead of the bowed-out type common to the productions of Vidler's factory. It wears, in consequence, a more elegant appearance than most coaches of that time; but it must be confessed that what it gained in the eyes of the passers-by it must have lost in the estimation of the insides, for the interior is not a little cramped by those straight sides. The guard's seat on the 'dickey'—or what in earlier times was more generally known as the 'backgammon-board'—remains, but his sheepskin or tiger-skin covering, to protect his legs from the cold, is gone. The trap-door into the hind boot can be seen. Through this the mails were thrust and the guard sat throughout the journey with his feet on it. Immediately in front of him were the spare bars, while above, in the still remaining case, reposed the indispensable blunderbuss. The original lamps in their reversible cases remain. There were four of them—one on either forequarter, and one on either side of the fore boot, while a smaller one hung from beneath the footboard, just above the wheelers. The guard had a small hand lamp of his own to aid him in sorting his small parcels. The door panels have apparently been repainted since the old days, for although they still keep the maroon colour characteristic of the mail-coaches, the Royal Arms are gone, and in their stead appears the script monogram in gold, V.R."

It is the coach which of all vehicles has least changed its appearance in the last hundred years. The drag of to-day and the old coach just described differ from one another only in a few minor details of construction. The reason for this is not far to seek. "The brief 'Golden Age,'" says Sir Walter Gilbey, "of fast coaching saw the vehicle, of which such hard and continuous work was required, brought as near perfection as human ingenuity and craftsmanship was capable of bringing it. No effort was spared to make the mail or road-coach the best possible conveyance of its kind, and in retaining the model of a former age the modern coachbuilder confesses his inability to improve upon the handiwork of his progenitors."

It is curious to note, by the way, that for a short time such coaches were hardly made at all, and the Report on the carriages shown at the London Exhibition of 1862 speaks of the "revival of an almost obsolete carriage, the four-in-hand coach, which had taken place within a few years." This was undoubtedly due to the founding in 1856 of the Four-in-Hand Driving Club.

Nor was this revival confined to England. In the official *Reports upon Carriages* at the Paris Exhibition of 1878, Mr. G. F. Budd draws attention to the fact that "the French have closely adhered to the English styles in the general design and shape of the bodies of their vehicles, especially in broughams ... landaus ... and drags. In the latter description of carriage, which has become so popular during the last few years, though it is peculiarly an English carriage, the style has been closely followed, and with such considerable success, that the French builders now appear as our formidable rivals in this branch of the manufacture." "A novelty," he continues, "in the design ... consists in the roof being so constructed as to admit of being opened in the centre ... a cover is placed on the top of the two portions of the head thus opened, and so forms, to all appearance, an ordinary luncheon-case with the ends open: it thus serves the purpose of a table when required ... and affords an increase of ventilation to those riding inside the vehicle." Similarly in America drags began to be built after the establishment of a driving club. These are identical with the English models.

With regard to the other four-wheeled carriages, we have now arrived at a period when it is almost impossible to speak at any length of each particular kind.[57]

---

[57] For full and particular accounts of all such carriages as have been constructed since the middle of last century, the reader is referred to the various trade journals. Further information is to be obtained from the Reports on carriages at the successive London and Paris Exhibitions. Here the more important differences between English, French, and Austrian carriages are clearly shown in a language which is not too technical for the ordinary reader to understand.

For in the first place such a classification as I have used to describe the older vehicles must to a large extent break down, and in the second place, from the time when the great exhibitions did so much to make the manufacturers of all nations familiar with each other's work, nearly every coachbuilder of standing has produced one model, if not more, peculiar to itself. So, in the middle of last century, you had carriages which approximated to the barouche, yet which had been evolved indirectly from so different a vehicle as the phaeton. You saw carriages, obviously dissimilar in appearance, yet bearing, to the layman, the same name. You had new combinations of perches and springs. And carriages were being exported from one country to be improved upon the lines most suitable to the roads and tastes of another.

Of all these carriages perhaps the two which deserve most mention are the *landau* and the *victoria*, both open carriages, which can be closed at will.

The *landau*, as I have said, had originally been a coach made to open. At the beginning of the century it had hardly been so popular as the *landaulet*, but at this time it underwent several improvements at the hands of Mr. Luke Hopkinson, a celebrated coachbuilder of Holborn. It was Hopkinson who first built what was known as a *briska-landau*, but he chiefly concerned himself not so much with the shape of the carriage-body as with the hood. He built his new landaus in such a way as to allow the hood to be folded, so that it lay horizontally at the back of the seat. At the same time the floor and the seats were raised so as to make the whole carriage a far more spacious and comfortable vehicle than had been possible when the hood could not be completely opened.[58] And with the hood entirely "down" you had practically the landau of to-day, possibly the commonest carriage on the road. Nearly every "fly" which so often is to be seen standing rather forlornly outside the village station as your train thunders past is a landau modelled on Hopkinson's designs. He was not, however, the only coachbuilder whose attention was being given to this useful carriage. Of one of the new landaus built by other firms a trade journal of the day observed with some truth that "its graceful outline and roominess" made it "the very beau-ideal of vehicular luxury." And as the years passed the landau in its several varieties increased in popularity. Improvements tended almost solely in the direction of lightness. The Report on the carriages at the exhibition of 1862 pays particular attention to the landau. "The demand for them," it runs, "has ... increased. They are well suited to the variable climate of the British Isles, as they can readily be changed from an open to a closed carriage and vice versa." At a later exhibition — in 1885 — the landau[59] had become

[58] This was also the case in France.

[59] There is an interesting passage in the 1878 Report which may be quoted here. "It is somewhat singular," this runs, "that while the attention of the English coachbuilders has, for the past few years, been directed to perfect an arrangement to open and close landau heads in a simple and effectual manner, the French builders have paid little or no heed to the attainment of this desideratum, but have instead adopted a plan which allows of the doors of a landau being opened when the glass is up, being first introduced by M. Kellner ... in 1866.... The simplest method is to have two pieces of brass, about ten inches long, in the form of a groove, for the glass frame to slide in, hinged to the upper extremities of the door pillars, and to close down on the fence rail when not required for use."

so popular that there was actually shown one, built for the Earl of Sefton, suited to the capabilities of a single horse. This was an important departure from tradition which seems to have shocked some of the old-fashioned designers. "That an established house with an aristocratic connection," lamented one trade paper at the time, "should exhibit a landau for one horse would have been considered incredible twenty years ago." No doubt this was true, but people persisted in their desire for light carriages, and a one-horse landau was the natural outcome. At a later date there was a tendency to alter the shape of the body. Hitherto this had generally been angular; now the lines became curving, the body, looked at from the side, forming the arc of a huge circle. Such a carriage was known as the *canoe landau*. To-day the canoe bodies, both in England and abroad, are made rather deeper than at the time of their introduction, but the square shape still persists. If there is one English vehicle which may be called the favourite carriage it is surely the landau.

Princess Victoria in her Pony Phaeton

(From a Drawing by Lowes Dickinson, 1835)

The earlier history of the *victoria*, the landau's chief rival, is rather obscure. As I have mentioned, the once popular *cab-phaeton* was still to be seen in the 'forties in many continental cities as the *milord*, which from a most aristocratic vehicle had descended into the realms of hackdom. An English coachbuilder, however, Mr. J. C. Cooper, saw possibilities in such a vehicle and prepared a series of designs. His drawings were scornfully treated in England, but "found favour in the eyes of his continental clients," who about 1845 constructed from them a four-wheeled cabriolet with seats for two. This small open carriage was copied in more than one place, particularly, it would seem, in Paris and Vienna. Whether these copies were still called *milords* I am not sure, but in 1856 they seem to have been described as *victorias*. In the meantime the pony phaeton had become popular in England, and in 1851 a new model designed for Her Majesty was, according to Stratton, also called a victoria.

> "In the summer of 1851," he writes, "a unique little pony phaeton was built by Mr. Andrews, of Southampton, for the Queen. The original announcement stated that when the carriage was delivered in front of the palace in the Isle of Wight, 'the Queen and Prince expressed to Mr. Andrews their entire satisfaction with the style, elegance, and extraordinary lightness and construction of the carriage,' which scarcely weighed three hundredweight. The height of the fore wheels is only eighteen inches, and of the hind ones thirty inches. The phaeton is cane-bodied, of George IV style, with movable head; the fore part is iron, but very light and elegant and beautifully painted. This carriage is known as the victoria, and has since been much improved in England and America."

Mr. Stratton is probably right; but it was the French-built carriage which the then Prince of Wales brought to England in 1869 to which the name may be more correctly ascribed. It is to be noticed, however, that the pony phaeton and the victoria proper differ from one another only in size and in the presence or absence of a driver's seat. The Prince of Wales's carriage was curved in shape and hooded, but about the same time Baron Rothschild imported a victoria from Vienna of the square shape. Both forms persist. At first, of course, the victoria was looked on with suspicion, but the Princess of Wales speedily showed her liking for it—it did indeed make an ideal lady's carriage—and in a short while the world followed suit. "Light, low, easy, fit for one horse, and looking very well behind a pair of cobs," remarks Thrupp, "it is not surprising that the victoria meets with so much patronage." At first it would seem that the hood was not made to lie flat, a fact amongst others which prompted a caustic critic in 1877 to grumble at the conservatism of English manufacturers. "Even with so good a model of this carriage as that presented to them in the victoria," he wrote, "the English builders do not see fit to maintain the same lines, and for some inscrutable reason deem that the hood when down should rest at an angle; whereas the 'cachet' of the Parisian equipages lies in the absolute straight line it maintains with the horizon." Only a few years later, however, another critic was drawing attention to the superiority of the English victoria over its French counterpart. "Their rattle," he wrote of the latter, "is enough to distinguish them. The French victoria is a low-mounted and decidedly unsymmetrical machine. The pole [is] a foot longer than it should be, the splinter bar and fore carriage too low"—a criticism which holds good to-day with most of the Italian carriages of this type.

Varieties of the victoria were constructed almost as soon as the carriage had reached to any degree of popularity. A hinge-seat was fitted into the front boot to face the ordinary seat, and this not proving enough, a permanent seat for two was built in its place, this innovation giving rise to the double victoria, which was built with or without doors. I need not, perhaps, dwell further on the victoria, except to observe that such changes as took place in the landau also took place in its more delicate rival.

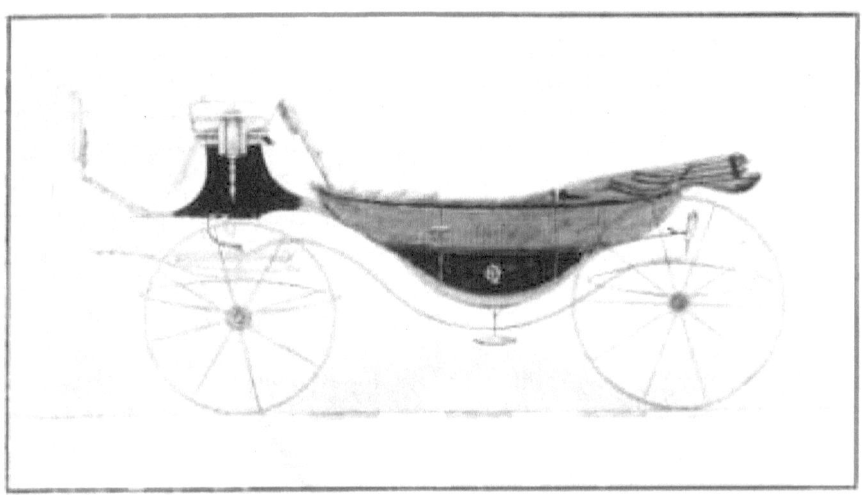

Canoe-shaped Landau, 1860

Drag, 1860

Another open carriage which remained popular until the introduction of automobiles is the phaeton. Sir Walter Gilbey mentions several varieties. Of these the largest seems to have been the mail phaeton.

"It was a favourite carriage," he writes, "seventy years ago or more, and was frequently used by gentlemen for long post-ing journeys in England and on the Continent. In these days this carriage was always built with a perch, the undercarriage resem-bling that of a coach, whence its name. For a time elliptical springs were adopted, but during the last ten years the fashionable mail phaeton has been a solid-looking square-bodied vehicle on its old undercarriage."

In 1889, he also observes that a jointed perch was used, the object being "to prevent the vehicle being twisted on bad roads, and also to preserve its equilibri-um under trying conditions of roads." The *demi mail phaeton*, to which Sir Walter gives the credit of having ousted the ugly *perch high phaeton* from public favour, "derives its names from the peculiar arrangement of the springs in the construc-tion of the undercarriage." Another variety, the *Beaufort phaeton*, is large enough to carry six people, and was, in the first place, expressly designed to carry people to the meet. Yet another modification, the *Stanhope phaeton*, invented by the peer of that name, is smaller than the last-mentioned, and has achieved a world-wide popularity. "The head and apron render it suitable for winter work, and when the hood is thrown back the stanhope is an admirable vehicle for summer use whether in town or country." The *T-cart* is a smaller *stanhope* "with compassed rail and sticked body in front and a seat for the groom behind." Sir Walter records the fact that its greatest popularity was about 1888, after which it was supplanted by the *spider phaeton*—a "tilbury body on four wheels with a small seat for the groom supported on branched irons behind."

It would be possible to mention half a dozen other varieties of the phaeton,[60] but such a list is best relegated to a coachbuilder's catalogue. There is only one innovation which should not be allowed to pass unnoticed here. Many of the pha-eton bodies during the 'sixties were constructed of basket-work; indeed, Croydon, where lived the inventor, received all the benefits which a new industry brings in its trail, but the popularity of these basket-carriages waned as rapidly as it had waxed—due, according to one writer, to the ridicule heaped upon them by *Punch*. A revival was attempted in 1886, and "we have a reminiscence of it in the imitation cane-work painted on the panels of many carriages" at a still later date.[61]

We come to the closed carriages.

The *brougham* was undergoing about as many changes and improvements as fell to the lot of any other carriage, yet superficially it maintained much the same appearance. The coupé brougham so popular to-day is the relic of the old chari-ot.[62] Of its several varieties the best-known is, or rather was—for it is rarely, if ever, seen now—the *clarence*. "It was introduced," says Sir Walter, "about the year

[60] Here, I suppose, should be included the *Eridge cart*, invented by Lord Abergaven-ny. It holds four persons on two parallel seats.

[61] The phaeton has found particular favour in France. At the Paris Exhibition in 1878 was shown a phaeton built at Rouen, which, according to the official Report, was "the finest small carriage exhibited in the French department for ingenuity and fitness for work."

[62] Sir Walter Gilbey had a *posting brougham* built for his own use, which to an even greater extent resembled the old chariot. In this case postilions were used.

1842 by Messrs. Laurie and Marner, of Oxford Street, and has fairly been described as "midway between a brougham and a coach." It had very curved and rather fanciful lines, seated four persons inside, and was entered by one step from the ground, carried the coachman and footman on a low driving seat, and was used with a lighter pair of horses than the family coach." Certain models, however, show the driver's seat to have been high, on a level, that is to say, with the roof; and not long after the first clarence was designed, Lytton Bulwer caused to be built what was called a *Surrey clarence*, which possessed a hammercloth. The attempt, however, to produce a miniature chariot did not succeed. Another variety, named uncomfortably the *dioropha*, was shown at the Great Exhibition of 1851.[63] Here the side windows would slide up and down upon a new principle, and "the whole upper part of the body from the elbow-line could be lifted from the lower, leaving a barouche body." You were shown models of this upper portion hanging rather forlornly from the roof of a coach-house. But improvements in the landau caused the extinction of the *dioropha*, which does not seem to have been built after 1875. The *amempton*, invented by a Mr. Kesterton, was a smaller form of this carriage. The one-horse "growler" or "four-wheeler," by the way, which still wanders up and down the streets of London, is the lineal descendant of the *clarence*.

Of the more unconventional four-wheeled carriages, the waggonette seems to have been introduced about 1845 by the Prince Consort after a German model, though one writer gives the credit of the design to the Prince himself. Here, as every one knows, the seats faced each other at right-angles to the driver's seat, the door being at the back. At first they were built very large—to carry out the original intention of providing a *family* carriage which should really be worthy of the name. Afterwards smaller models were produced, and proved equally popular. "The principle of riding sideways," remarks Thrupp, "was not new. The Irish car, the four-wheeled Inside car of the Westmorland district, the old Break, and the Omnibus all contributed to the design of the modern vehicle." A few particular varieties may be mentioned. The now forgotten *perithron*, a Suffolk invention, was a waggonette in which the driving seat was bisected down its centre, so as to allow a passenger entering from the back to reach the front seat. The *Portland waggonette*, built for the Duke of Portland in 1893, was a large carriage with a folding hood. Another carriage of the kind with a folding leather hood was presented by Lord Lonsdale to the King and Queen at the time of their wedding. This is known as a *Lonsdale waggonette*. "Lord Lonsdale," remarks Sir Walter Gilbey, "allowed his name to be given to this device under the impression that he was the first to originate a head of this description; but his claim for invention of it was disputed at the time. Mr. Robertson stated that he had built such a waggonette so far back as 1864; Mr. Kinder had built one in 1865; and Messrs. Morgan stated that they had turned

[63] "The Patent Dioropha, or two-headed carriage, combining in one a clarence or pilentum coach, complete with all its appointments; a barouche, with folding head and three-fold knee-flap; and an open carriage. The heads can be removed or exchanged with facility by means of a pulley attached to the ceiling of the coach-house, aided by a counterpoise weight." *Vide* the Official Catalogue, which also gives illustrations of several Indian carriages, such as the *Keron*, the rath, a Mahratta carriage from Bengal, and a lady's carriage from Lahore—the last being a four-wheeled conveyance covered with scarlet and crimson cloth, and shut in with thick curtains.

out a similar vehicle before the year 1870." A very large waggonette, the *brake*, is a common enough object to-day, and is built in various forms. Sometimes a second seat is placed directly behind and parallel to the driver's seat. In some models these seats stretch back throughout the length of the carriage, in which case it is a *char-à-banc*. Awnings, permanent or temporary, are generally provided.

In America the commonest four-wheeled carriage is the light *wagon* or *buggy*, a name given in England to a light two-wheeled, single-seated cart (also called a sulky[64]) towards the end of the eighteenth century. The *buggy* has one seat fixed on to a long, shallow tray; the *wagon* is similar, but has two or more seats.

"These American waggons," says Thrupp, "were modelled from the old German waggon, but they have been so much improved as to be scarcely recognised. The distinctive feature of the German waggon was a light, shallow tray, suspended above a slight perch carriage on two grasshopper springs placed horizontally and parallel with and above the front and hind axle-tree; on the tray one or two seats were placed, the whole was light and inexpensive, and well adapted to a new, rough country without good roads. These waggons may still be found in Germany and Switzerland....

"American ingenuity was lavished upon these waggons, and they have arrived at a marvel of perfection in lightness. The two grasshopper springs have been replaced with two elliptical springs. The perch, axle-trees, and carriage timbers have been reduced to thin sticks. The four wheels are made so slender as to resemble a spider's web; in their construction of the wheels the principle of the patent rim used in England in 1790 has been adopted. Instead of five, six, or seven felloes to each wheel, there are only two, of oak or hickory wood, bent to the shape by steam. The ironwork of the American buggy is very slender, yet composed of many pieces, and, in order to reduce the cost, these pieces of iron are mostly cast, not forged, of a sort of iron less brittle than our cast iron.... The weight of the whole waggon is so small that one man can lift it upon its wheels again if accidentally upset, and two persons of ordinary strength can raise it easily from the ground. The four wheels are nearly of the same height, and the body is suspended centrally between them. There are no futchels; the pole or shafts are attached to the front axle-tree bed, and the front of the pole is carried by the horses just as they carry the shafts; the splinter-bar and whipple-trees are attached to the pole on swivels. Some are made without hoods and some with hoods. These are made so that the leather of the sides can be taken off and rolled up, and the back leather removed, rolled, or fixed at the bottom, a few inches away from the back, the roof remaining

---

[64] The only sulky now to be seen in this country is the trotting carriage used in races—à mere skeleton. See also p. 129.

as a sunshade....

"The perfection to which the American buggy or waggon has been carried, and every part likely to give way carefully strengthened, is marvellous. Those made by the best builders will last a long time without repair. The whole is so slender and elastic that it 'gives'—to use a trade term—and recovers itself at any obstacle. The defect in English eyes of these carriages consists in the difficulty of getting in or out by reason of the height of the front wheel, and its proximity to the hind wheel—it is often necessary partly to lock round the wheel to allow of easy entrance. There is also a tremulous motion on a hard road which is not always agreeable. It is not surprising that, with the great advantages of extreme lightness, ease, and durability, and with lofty wheels, the American waggons travel with facility over very rough roads, and there is a great demand for them in our colonies. It must be remembered that the price is small, less than the price of our gigs and four-wheeled dog-carts."

Modern American Station Wagon

Indeed, the tourist in America will come away with the impression that there is hardly a family in the continent which does not possess at least one buggy or waggon. They can be driven, too, at a very great pace. In this connection it is interesting to notice that it was a buggy which Lord Lonsdale selected in order to carry out his great driving feat in 1891, when "he undertook to drive four stages of five miles within an hour, using for the first three stages one, a pair, a team, and riding postilion in the fourth."

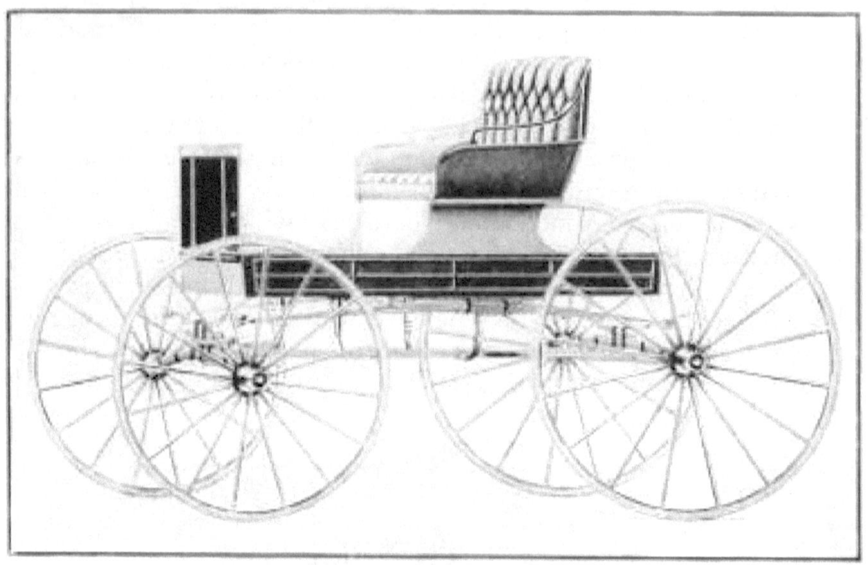

Modern American Buggy

Both from Studebaker's (Chicago) Catalogue

There are, of course, many varieties, several invented after Thrupp wrote the above account. Of these some are peculiar to a particular State, while others seem to be in general use throughout the continent. In Chicago, for instance, and other towns of the middle west, the commonest buggy seems to be the *bike wagon*, of which a variety is the *cut-under bike wagon*, where the tray is double—the seat forming a bridge between its two parts.

Stanhopes and phaetons are also manufactured in America, though on a much lighter scale than in England. Another popular American carriage is the *surrey*, which has the two-seated arrangements of the larger waggons, but is without the tray. The *station wagon*, very popular in New England, resembles the old English chariot, and differs from it only in its driving seat, which is on a level with the inside seat and directly against the front lines of the carriage-body. This is one of the most comfortable carriages in the country. The *buckboard*, even slenderer than the buggy, is hardly more than the skeleton of a carriage, but seems none the less popular on that account. The *barge* is the name given in Massachusetts to a two-seated waggon, and the word has a curious origin. It seems probable that it is a relic of the days when in that part of the country the boat sleighs used in the winter were put upon wheels in the summer. At a later date ordinary waggons were used for summer traffic, but the old name stuck. And I dare say there are a dozen or more local names of some peculiarity in other parts of America which to-day are given to carriages not in the least like those to which the name was originally applied.

Coming to the two-wheeled carriages, we find similar changes to those described above showing themselves. The old curricle, for instance, is now but rarely seen, its place being taken by one or other of the dog-carts. What was probably the most fashionable of these carriages during the early Victorian era is now practically extinct. This was the *cabriolet*, rather different in appearance from the vehicles of that name which had plied for hire but a few years before, yet built on the same principles as the earliest French gigs.

"They were greatly improved," wrote Mr. G. N. Hooper in 1899,[65] "about fifty years ago by the well-known Count D'Orsay and the late Mr. Charles B. Courtney, who greatly refined the outlines and proportions, making them lighter, more compact, and far more stylish. They became *par excellence* the equipage of the *jeune noblesse*, and no more stylish two-wheel carriages for one horse were driven for many years while they were fashionable. A large, well-bred horse was a necessity, and this the cabriolet generally had.

"The groom, or 'tiger' as he was then called, was a special London product: he was produced in no other city, British or foreign; all the genuine tigers hailed from London. His age varied from fifteen to twenty-five. Few there were that were not perfect masters of their horses, were they never so big. In shape and make he was a man in miniature, his proportions perfect, his figure erect and somewhat defiant: his coat fitted as if it had been moulded on him; his white buckskin breeches were spotless; his top-boots perfection; his hat, with its narrow binding of gold or silver lace, and brims looped up with gold or silver cord, brilliant with brushing, was worn jauntily. As he stood at his horse's head, ready to receive his noble master, you might expect him to say, 'My master is a duke, and I am responsible for his safety.'"

There is little enough to say of the gigs. The curricle, as I have said, is now rarely seen, though Sir Walter Gilbey mentions a particular one introduced about 1883 "which differed materially from the vehicle formerly known by that name. It consisted of a cabriolet, or whisky body, having an 'ogee' or chair back, the body being suspended by braces from C or S springs upon the undercarriage. Its peculiarity lay in the use of long lancewood shafts, set so far apart that the pole could be placed between them; the saddle-bar being used to support the pole, the shafts, it would seem, were somewhat unnecessary." The *Cape cart* brought into England from South Africa is a two-wheeled vehicle of this class with a pole in place of shafts, and "the sides being framed so as to present three panels."

"At the back," says Sir Walter, "was built in a large box for provisions, the full width and depth of the cart, the back seat forming the lid; the tail-board was used only as a foot-rest. An

[65] *Suspension of Road Carriages*. A Paper read before the Institute of British Carriage Manufacturers at York. 1899.

adjustable centre seat with backrest could be used so as to provide
accommodation for six passengers. A white canvas tilt on wooden
hoops with sunblinds at the sides, which could be strapped up
when not wanted, covered the whole body of the cart."

And similar to the Cape cart is the *Whitechapel cart*, which brings me to a brief
consideration of the *dog-carts*.

As originally designed, the dog-cart seems to have been built high, and, as its
name implies, for the purpose of carrying dogs. Such a vehicle would seat four, a
roomy, comfortable trap "with space under the seats, where a brace of pointers or
other dogs could lie at ease." As I have said in a preceding chapter, the sides of the
cart "were made with Venetian slats to provide ventilation." Such a cart, however,
proved so agreeable that no long time elapsed before its original purpose was lost
sight of, and it became one of the commonest of country carriages. Built on a small
scale it was admirably suited for pony or cob. Numerous varieties exist. In the *tandem cart*, as generally constructed, the driver's seat is high—the only cart, indeed,
of the kind to maintain any height at all. In the *Ralli cart* two seats are placed back
to back, the foot-rest to the latter closing on the body when required. (Built somewhat on the lines of the ralli, by the way, is the Indian *tonga*, "a rather low, hooded
vehicle ... furnished for draught by a pair of ponies on the curricle principle with
pole and bar.") The *Battlesden, Bedford,* and *Malvern carts* are other varieties. More
popular, perhaps, than any of these is the *governess cart*, which, while really in a
class by itself, may be mentioned here. This is a low and particularly safe carriage,
in which the seats are placed at the sides, as in the waggonette, and the door is at
the back. An improvement on the governess cart, though not nearly so popular, is
the *Princess car*, first designed in 1893. Here the back door is dispensed with, the
entrance being in front. "The driving seat is arranged on a slide, whereby it can be
moved forwards or backwards to adjust the balance; and it also enables the driver
to sit facing the horse instead of sitting sideways as in the governess cart."

In the last chapter I pointed out the chief varieties of public carriages. Of these
the hansom and the omnibus have undergone considerable changes. The hansom
was enormously improved by Mr. Forde, a Wolverhampton coachbuilder, in 1873,
when the Society of Arts offered a prize for the best two-wheeled public conveyance. Mr. Forde's carriage was much lighter than the older hansoms, and "its merits attracted the appreciative attention of foreigners, whereby an export trade became established." Four years later another vehicle, the *two-wheeled brougham*, was
introduced, but did not meet with success. The *Floyd hansom* of 1885 showed other
improvements, and for the first time the hansom became a private carriage. Here
the "side windows were made to open, as were two small windows at the back of
the cab." For a short while, indeed, the private hansom was one of the smartest of
gentlemen's carriages. Then in 1889 was shown another hansom with a movable
hood. This was wholly unsuccessful, but the *Arlington cab*, a Dorchester invention
of this time, may still be seen in provincial towns to which the taximeter petrol cab
has not yet reached. The chief peculiarity about this hansom is its doors, which,
instead of reaching only half-way up and being constructed at a backward angle,

Lector House believes that a society develops through a two-fold approach of continuous learning and adaptation, which is derived from the study of classic literary works spread across the historic timeline of literature records. Therefore, we aim at reviving, repairing and redeveloping all those inaccessible or damaged but historically as well as culturally important literature across subjects so that the future generations may have an opportunity to study and learn from past works to embark upon a journey of creating a better future.

This book is a result of an effort made by Lector House towards making a contribution to the preservation and repair of original ancient works which might hold historical significance to the approach of continuous learning across subjects.

HAPPY READING & LEARNING!

LECTOR HOUSE LLP
E-MAIL: lectorpublishing@gmail.com

in. The latter were ready for any adventure and available to any crime in order to give a justification for their miserable existence. Inserted in a culture genetically predisposed to violence and abuse, they were a disastrous destabilizing element, a dramatic problem for those communities that were trying to build a civil and livable society in the New World.

Garibaldi immediately gets himself into trouble because after a few months he receives an expulsion decree.

It is at this time that he enrolled in the Freemasonry and, perhaps, comes in contact with the English navy.

He will never write about this in his *Memorie* because this information must remain secret. Freemasonry and England will be decisive for his triumphs and for the creation of his myth. In those years the Perfidious Albion was fishing in troubled waters to destabilize the area, and it is likely that she favored Garibaldi for piracy actions. Freemasonry served as a link.

This expulsion is not a problem for him. Shortly before his arrival a revolt had broken out in southern Brazil that started the Revolução Farroupilha (the revolution of the rags). The province of Rio Grande do Sul tries to break away from the Brazilian Empire under the leadership of Bento Goncalves who declares the republic and appoints himself President. His secretary is an Italian fugitive, Luigi Zambeccari; they are all Freemasons.

They were talking about republic and liberty and Garibaldi takes fire immediately; the opportunity he was waiting for had finally arrived. He joins the rebels and for six years will fight for the "freedom" of the Republic of the Rio Grande do Sul. Problem is that freedom had nothing to do with this, absolutely nothing. Goncalves was one of those many fake liberators of which South America has always been so rich. As the same Garibaldi will notice later, the people of the Rio Grande will end up hating Goncalves but, as we shall see, this was absolutely irrelevant to Garibaldi. "*Apparently all exhibited a passionate love for freedom, but it was an anarchic freedom, which easily faded into partisanship and then into ruthless dictatorship*".[3]

What was it that fascinated Garibaldi in this enterprise? As he writes to us, it was the love of war: "*The life we were living in that class of war,*

---

[3] Denis Mack Smith, *Garibaldi* (Milano: Laterza, 1956) pag.16

was very active, full of dangers ... but at the same time beautiful, and very conforming to my character so inclined to adventures".

It may be that also the character of Goncalves had its influence. Thus he describes him: "*Bento Goncalves was the kind of brilliant warrior and magnanimous ... Top of the stature and quick, he bestrode a fiery steed with the ease and dexterity of a young man ... Sober like any child of that valiant nation. ... A most valiant person, he would fight in single combat, and perhaps win, any strong rider. On the soul generous and modest, I believe he did not excite the people of Rio Grande to emancipate themselves from the empire, for the purpose of his own aggrandizement".*

This is a grotesque misrepresentation of reality. Goncalves was a landowner and slaveholder and had started the secession from the Empire, along with other landowners, to become masters of the whole province and be able to carry on their business without having to account to any imperial officer. It is only a slight exaggeration to define him a bloodthirsty beast. A doubt arises; did Garibaldi really believe what he was writing? Or is this another case of his self delusional frame of mind?

But if you read carefully this description you can see that Garibaldi is describing himself. This was how he saw himself, or rather; this was what he wanted to be.

In fact this is the character that the myth has managed to fabricate.

In fact this is what the world thinks of him after two centuries.

Garibaldi and Rossetti armed a boat that they christened *Mazzini.* Hence they obtain a Letter of Marque by the new republic. The letter of marque was the authorization to plunder enemy ships while maintaining a base in the country which granted the letter. For this country it was a very cheap way to make war since the cost of the armament was borne by the corsair and his were the risks. It was also a very cruel war: the only victims were civilians.

Garibaldi goes to sea with 12 companions, almost all Italian exiles, and with an enthusiasm worthy of a better cause, "*Corsair! Launched on the ocean with twelve companions on board of a garopera, we were challenging an Empire, and we were the first to fly a flag of emancipation! The Republican flag of the Rio Grande".*

He was a strange pirate. One day he captures a boat loaded with coffee and with several civilians on board, but: "*not all my mates were ... men of pure costumes; and a few of them were making themselves grim to intimidate the innocent our enemies. I tried to repress them and to diminish the fear of our prisoners*". Maybe that's why it happened that "... *a Brazilian passenger came to me pleading and offering me three valuable diamonds. I refused them to him, as I commanded not to touch the individual effects of the crew and passengers. I kept this attitude in every similar circumstance and my orders were never transgressed being my subordinates sure, without a doubt, that I was not willing to compromise on this matter*".

No pirate in the world had ever done such a thing! That was absolutely extraordinary. The fame of his exploits spreads throughout South America, bounces off to Europe and the people start talking about this pure and selfless warrior who fights just for the freedom of the people. His myth is getting started. In France, Dumas jumps at the opportunity and begins to fabricate this character. Garibaldi did not know it yet but in Europe he was becoming famous. Perhaps, it is with Garibaldi that progressive Europeans begin to idealize the guerrillas of South America.

Now we must pause and ask ourselves: was it true? Did Garibaldi really behave like this? After all he and his pirates lived of robberies; is it possible that his pirates kept following him at these conditions? How would they get rich?

That was probably true. All the following history proves that Garibaldi had a charisma and an authority strong enough to easily impose himself to the bands of wretches who followed him. This did not always happen, they still remained a gang of thugs but, sometimes, they consented to satisfy their leader. Perhaps they too did not mind making the figure of heroes rather than that of pirates, sometimes.

In this case we must specify that the passenger with the diamonds was a Brazilian who had sold all his property and invested everything in coffee to move and change his life; he was carrying 428 bags of coffee and four slaves. We can assume that he was grateful to Garibaldi for having left him his life and the diamonds when, after a few days, he was landed on a desert beach, in misery but unharmed. Normally these corsairs cut the throat to the prisoners that could not be sold as slaves.

Now we can begin to trace the path of our character to try to figure out what kind of person he was.

He participates in subversive activities in Genoa and the Kingdom of Sardinia condemns him to death. He repairs in Brazil where the Empire gives him hospitality and his compatriots offer him the opportunity to rebuild his life with what he can do: the sailor. Unfortunately this trade goes wrong; therefore his passion for war takes over and he associates himself with Freemasonry a powerful society more or less secret and he begins operations of piracy and/or smuggling using, it seems, the boat that, it seems in good faith, was given to him for business operations. The Empire reacts and expels him. These latest events are never mentioned by him. The secession of the Rio Grande do Sul from the Empire gives him the opportunity to make war by granting him a letter of marque so that he is, officially, a freedom fighter and not a criminal. Here he resumes his *Memorie*. He goes to fight for the secessionists without even wondering if it was the right thing to do. He exalts his comrades-in-arms with a brazenness that goes beyond the ridiculous. In his "military" operations he strives to keep to a minimum the damage to civilians. It must be clear to the public that, even if he has to kill and destroy, he does not forget his love for the people. His charisma is huge: blond hair, holy eyes, a good face, plays with children, intelligent, strong and brave. His men obey him and they do well, those who follow him to the end will be brought to Italy to participate to the Risorgimento and today a few streets in Italy bear their names.

We cannot but conclude that Garibaldi was a genius of public relations and … a criminal.

We continue reading his *Memorie*: "It (the war) *was not limited just to the navy. We had saddles on board; horses we found everywhere in those countries where they are abundant; and all at once, when the case required it, we were transformed into a not brilliant but fearful and feared cavalry. There were on the shores of the lagoon the estancias, which the events of the war forced their owners to leave. There you could find cattle of every kind to eat and to ride"*. What a wonderful adventure! Accommodation, food and horses free, giving bottom to the re-

sources abandoned by the owners who had been forced to flee their homes!

Who were his comrades? "*The people that accompanied me were a truly cosmopolitan crew, made up of everything and of all colors, as of all nations. The Americans for the most part were freedmen, blacks or mulattos, and generally the best and most trusted. Among the Europeans, I had the Italians including my friend Luigi and Edoardo Mutru, a friend from childhood, seven in all whom I could count on. The rest was made up of that class of adventurer sailors known on the American shores of the Atlantic and the Pacific under the name of "Frères de la cote". A class that had certainly provided the crews of pirates, buccaneers and today still gave its contingent to the trafficking of blacks*".

It should be noted that among his men there are no people from Rio Grande; evidently the people did not side for the secession. There are the Negro slaves of the landowners enrolled by their owners with the promise of freedom; at the end of the war, if winners, they would have been freed (perhaps!). And they were the most reliable since the others were criminals. The Italian exiles, all of them had something to settle with justice (as Garibaldi himself) and then the "Brothers of the Coast". Pirates on the run, desperate people ready for anything, veterans from the slave trade that a few years before had been outlawed by England. Who knows how they felt, the Negro slaves, to fight alongside their tormentors. Not surprisingly, all the inhabitants of those "liberated" lands had fled. Is it possible that it did not come to mind to Garibaldi any doubt about the goodness of the cause for which he was fighting? If the people had fled and he was reduced to command a band of thugs and desperadoes, if no people from Rio Grande had joined them, maybe his cause was not the right one! **What a folly!**

Another question arises: the fan clubs of Garibaldi, did they ever read his *Memorie*?

His friend Rossetti had left the armed struggle to devote himself to propaganda. A printed magazine, *O Povo* (The People), was established in the capital of the insurgents. Evidently, in spite of what Garibaldi just told us, they pretended to represent "The People". These propaganda initiatives are very important because they spread out to South America and arrive in Europe. They are avidly read by European progressives and produce the hard core of the myth of Garibaldi. They are also a magnet

for the European exiles who want to fight for the Republican cause. The man who directed the works to build the boats of Garibaldi was of Irish origin, John Griggs.

This adventure of Garibaldi seems to be taken from *Gone with the Wind*: "*... they lived for the extension of most of the river, stretching over a vast area, all the families of President Bento Gonçales and his brothers, The estancias where we landed were those of Donna Antonia and Donna Ana, both sisters of Bento Gonçales ... I can assure you that none of the circumstances of my life comes to my mind with more charm, more sweetness and more gentle reminiscence of that which I spent in the lovely company of those ladies and their dear families*". What an accomplished gentleman!

One of these estancias housed three sisters: "*... and one of them, Manuela, reigned absolutely over my soul. I never stopped loving her, though without hope, since she was engaged to a president's son. I loved the ideal beauty of that angelic creature and my love had nothing profane. On the occasion of a fight, where I had been presumed dead, I knew I was not indifferent to that angelic creature; and that was enough to console me for the impossibility to possess her*". How romantic!

He continues: "*Also the slaves of color were not indifferent, who were housed in those elegant establishments, and they could be adored with a cult a little less than divine*"[4]. What a wretch! How can you write words of such a passionate love and just after two lines inform the reader that while he was pining for love of an angelic creature, he was reaping hits with the slaves?

We should not be surprised if the slaves were not indifferent to the charms of the blond hero. In fact, if his charisma was magnetic on men, on women it was devastating. Garibaldi was the man who never had to ask ... and who never said no. It is not an exaggeration to say that women rained on him. He took all of them, ugly or beautiful, aristocratic or plebeian, rich or poor, nubile or married.

It should not surprise you that the masters of the estancias let him have his way. By law, the children of the slaves were slaves anyway, whoever was the father, and belonged to the owner of the slave. With

---

[4] Francesco Carrano, *I Cacciatori delle Alpi comandati dal generale Giuseppe Garibaldi* (Torino, Unione Tipografica Editrice, 1860) pag. 35. We had to consult this old document since this sentence is censored in the edition of the Memorie of BUR, which is very incorrect.

this law the owners could have fun with their properties and earn on the children that were born. In fact, the sale of the children of the slaves was a very important asset for the landowners who were in difficult financial situations. The slave trade from Africa had been outlawed by the English empire in 1833 and this had increased the price of the slaves in America. The Brazilian Empire abolished slavery in 1871 with the Law of the Free Womb: all born to the slaves should be considered free. By means of this law, slavery disappears gradually without disrupting the society, as it happened in the United States.

We are sure that both the slaves and their owners were satisfied with the performance of our Hero of Two Worlds. This dichotomy between the woman "angelic creature" and the woman "object of lust" is typical of Romanticism and therefore should be considered normal for the times. It is not normal at all to write it in such a brazen way. How to interpret this effrontery of Garibaldi? An excess of sincerity, another inconsistency or the arrogance of he who knows he can do anything he wants because those who love him will love him anyway?

As we have already said, his mental inconsistencies were authentic and worsened with time. This, in particular, will put him in a dramatic situation when after the first war of independence he marries the young Countess Raimondi (another angelic creature: she 18 he 53 years old!) to be informed, right after the marriage, that his wife was pregnant by a stranger. His efforts to obtain the annulment of the marriage will go on for years giving him terrible headaches.

However, we cannot fail to add that, even though we do not sympathize with our hero, in reading his memoirs we cannot help being fascinated by his adventures. We can sense the charm of his life and the impact it will have on his contemporaries. Who knows how many young patriots have dreamed of riding in the endless pampas, battling sinister tyrants and courting romantic damsels. Who knows how many languid maidens have dreamed of being possessed by some blond hero.

Sooner or later the dream had to end, luck turns its back on him and he shipwrecks: *"The survivors, numbering fourteen, one after the other all of them had landed. In vain, among them, I looked for an Italian face. All dead! It seemed that I had been left alone in the world! I wandered, and it almost seemed heavy to me this existence saved with great effort"*. The death of all his fellow Italians provokes an existential crisis. He had

never thought of marriage because "*having a woman, children, seemed to me something entirely unseemly for someone who was totally devoted to a principle* (the redemption of the Motherland)", but now he must recognize that "*fate had decided otherwise. I, with the loss of Luigi, Edoardo and my other fellow countrymen, had been left in a bleak isolation. I felt I had been abandoned alone in the world*". He concludes that marriage was the only remedy to this crisis.

As he pondered these considerations he was moored in the harbor of a small town they had just conquered: Laguna. "*I was walking on the quarterdeck of Itaparica enveloped in my gloomy thoughts; and after considerations of every kind, I concluded at last to look for a woman that could draw me from this boring and unbearable condition. I threw at random my gaze up at the homes of the Barra ... There, with the help of the telescope, that usually I held in hand when on the quarterdeck of a ship, I discovered a young woman. I ordered to be carried to the shore in her direction. I landed; and going towards the houses where was to be the object of my journey, I could not find her: there I met with an individual of the place whom I had met the first time of our arrival. He invited me to have a coffee in his house. We entered: and the first person that appeared in front of my eyes was the one whose appearance made me land.*

She was Anita! The mother of my children! The woman, whose courage I often wished for me! We were both ecstatic, and silent, looking at each other like two people who do not meet for the first time, and seek in the features of each other something that facilitates reminiscence. I greeted her at last, and I said: you must be mine. I spoke too little Portuguese, and I articulated those impudent words in Italian. Anyway, I was magnetic in my insolence. I had tied a knot, I passed a judgment that only death could break! ... I had met a forbidden treasure, but also a treasure of great price! If there was fault, I had it all! And ... there was fault! Yes ... two hearts were tied with immense love, and the existence of an innocent was broken! ... She's dead![5] *I unhappy! And he avenged ... Yes avenged! ... I erred greatly and I erred alone!*".

Let's recap this soap opera: Garibaldi is in an existential crisis and reconsiders his views on marriage, concludes that the solution to his crisis

---

[5] Anita was dead when he wrote these memories.

is to get married, he needs a woman. Said and done, he grabs a telescope and finds the right woman. Just think, love at first sight at the telescope! He goes on search but does not find her. By chance an acquaintance invites him to his home and there she is, the wife of this acquaintance. With no respect for her husband he tells her that he will make her his woman. Here the story stops and he tells us about his guilt; what guilt? What does it mean that "*the existence of an innocent was broken*"? I think it should be taken in the literal sense: that Garibaldi, or both, murdered her husband.

Anita had married three years earlier, at 15, the shoemaker Manuel Duarte who was much older than her. They had no children and lived a very modest life. It is not difficult to suppose that she was very sensitive to the charm of a leader of the rebels that just conquered the town. We do not know the details of this operation but from then on Anita will always be at the side of Garibaldi and will dedicate to him her entire life, literally. The two start a rumor that Duarte, humiliated by Garibaldi, had joined the imperial army and then died at war somewhere. However, when a few years later the two decide to get married, in Montevideo, the priest will ask the proof that Duarte was dead; otherwise he would have not married them. At this point Garibaldi gives the proof of Duarte's death because he had killed him. Garibaldi was the leader of the pirates and rebels who had just conquered the town, he could easily get killed anybody he wanted. After all, he was General Garibaldi and the poor Duarte was a nobody.

Let's review this story at the light of what we know. Garibaldi knows he has a need, identifies the solution, finds it and then he takes it, someone opposes, he kills him. Many years later, in writing his memoirs, he puts together a romantic melodrama were he does not states clearly that he killed him, but he is so sorry that the reader cannot help but sympathize with him, not with the victim. All this is communicated with a wonderful prose. There is no doubt, he is a genius!

Now, however, we have to go back on his memoirs, to analyze his ideas about marriage: "*having a woman, children seemed to me something entirely unseemly for someone who was absolutely dedicated to a principle, that although excellent, would not allow me, who was dedicating myself with all the fervor of which I was capable, the peace and sta-*

bility necessary to a family man". This is how he saw himself, the High Priest "*absolutely dedicated*" to the Cult of the Nation. To it he wanted to devote himself entirely with no liaisons to women (apart from the physiological needs) and children; therefore, no family, like the Catholic priests. He wanted to be an ascetic of nationalism but "*the fate decided otherwise*"; he gives in and does something which is very natural. This is a very important element of his thought because asceticism is a necessary step towards criminal folly. However, his concern for "*the peace and stability necessary to a family man*" offers a glimpse of the son of two good Catholic parents. He was right to be concerned: he'll be a bad father.

After leaving Laguna, he resumes the war at sea but the situation is getting heavier for the rebels; evidently the imperial army has learned the lesson and begins to reap the first successes. The consequence is that the population takes courage and starts to show its hostility towards the rebels and its sympathy to the empire.

At this point the civil war becomes more brutal. The revolutionaries occupy the province of Santa Caterina: "*... but our arrogant demeanor towards the good people of Santa Catarina, our friends at the beginning but absolutely hostile in the end ...*" makes that someone rebels against the republic. "*With the advancing enemy, very big on land, and the overbearing demeanor, with which we treated the people of Santa Catarina, led some people to raise against the republic, and among others the town of Imiriù .... General Canabarro gave me the odious task of subduing the town and as a punishment to plunder it. I was obligated to fulfill the command. And even under a republican government it is clearly repulsive to have to blindly obey ... I wish for myself, and anyone who has not forgotten to be a man, not to be forced to give sack. I think that it is impossible ... to narrate minutely all filthiness and wickedness. I've never had a day of so much regret, and so much nausea of the human family*" Garibaldi tries to limit the damage, but in the town there was a deposit of alcohol. His men get drunk and his authority cannot make them stop. "*Finally, with threats, beatings and killings we managed to embark those wild beasts*". And even this cannot shake Garibaldi from his faith in the cause.